The Chemistry and Physiology of Growth

The Chemistry and Physiology of Growth

J. H. NORTHROP
F. O. SCHMITT
K. V. THIMANN
KARL FOLKERS
C. B. VAN NIEL
E. S. G. BARRON
P. WEISS
J. S. NICHOLAS
C. P. RHOADS
C. N. H. LONG

EDITED BY ARTHUR K. PARPART

KENNIKAT PRESS
Port Washington, N. Y./London

THE CHEMISTRY AND PHYSIOLOGY OF GROWTH

Reissued in 1972 by Kennikat Press by arrangement
Library of Congress Catalog Card No: 79-159099
ISBN 0-8046-1642-6

Manufactured by Taylor Publishing Company Dallas, Texas

FOREWORD

An architect must build upon the past. A study of the growth of animal and plant organisms must start from a firm foundation in the mechanics of growth which finds its most eloquent expression in D'Arcy Thompson's *Growth and Form.*

The symmetrical groupings which cells take in forming parts of organisms, the time required for the attainment of such groups, the numerous internal and external forces that give form to the individual cells, the symmetry of form of the whole organism and between organisms, all these and numerous other precisely measured parameters are the foundations upon which the students of growth must build.

The individual chapters of this book represent a number of sturdy columns that are being erected upon these foundations. Examined individually they make the reader want to take part in their construction. Viewed as a whole they present a hope for a future understanding of the phenomenon of growth.

One of the fundamental processes in growth of plant or animal consists in the manufacture of an ever-increasing amount and variety of proteins. These proteins form the major structural components as well as the organic catalysts of all living cells. Hence an analysis of the mechanisms involved in the "Synthesis of Proteins" (Chapter I) is an important part of a study of growth.

"Morphology at the Molecular Level" (Chapter II) is rapidly becoming an essential aspect of growth. It has already contributed greatly to our understanding of the problem of the arrangement of protein and other molecules within a cell, which determines the cells highly specific form, symmetry and reactivity.

At this point a question arises. What determines the mass and direction which these structural units and their accompanying water and salts acquire in the growing organism? The problems raised by this question differ somewhat in the growth of plant from that of animal organisms. For this reason Chapters III and IV deal with "Plant Growth Hormones" and "Vitamins and Growth Factors."

Organisms, whether unicellular or multicellular, are composed of cells which have arisen from preexisting cells. This is the fundamental biological phenomenon of growth. Our present knowledge and future prospects of understanding this phenomenon have been greatly accelerated in recent years by studies on the "Kinetics of Growth of Microorganisms" (Chapter V). This leads naturally to a desire to know more about the energy sources for the growth processes. What starts and maintains the chemical reactions that bring about the increase in mass, the elaboration of new protoplasm? Up to this point we are in the position of a man who has the plans for a building, all the many types of material necessary to construct it, and many willing hands to do the labor, but has failed to provide the foreman whose job it is to guide, to direct, to speed up or slow down as occasion demands, the action of those hands on the material available. In many respects the enzymes of living organisms are analogous to these foremen. In the case of enzymes the action is on the chemical reactions going on within cells. "Cellular Metabolism and Growth" (Chapter VI) brings out many important aspects of this problem.

The first six chapters of the book appraise many of the basic physiological mechanisms and chemical reactions that confer on the individual cell its form and function in growth. In the development of the multicellular organism, however, these units of form and function appear as part of the diverse types of structure that compose the adult organism. The remaining chapters of the book suggest how to deal with these problems.

Among the central problems of growth are those concerned with differentiation of cells. How is it that in embryonic growth, and during the various types of growth which occur in an adult organism, cells of similar origin come to possess widely divergent characteristics? What controls different modes and rates of growth? What are the factors which govern the establishment of biochemical and morphological differences among cells? These are the problems of "Differential Growth" (Chapter VII).

The differentiation of the developing egg as it goes through embryonic stages to the adult condition presents many "Problems of Organization" (Chapter VIII). In some studies it has been shown that there are localized regions of the egg, in some cases visibly different, which determine the pattern of development. If these areas are disoriented, visible alterations in the normal development of the embryo occur. In other studies on the early stages of development of the embryo specific groups of cells influence the activities of neighboring cells in such a way that they determine the fate of their neighbors.

Growth does not always proceed in accordance with the functional needs of the growing or adult organism. All too commonly the cells of a particular tissue or organ will go on a growing spree of their own. Such abnormal growths, neoplasms, are a part of the overall problem of growth. Examination of them should lead to a mutually stimulating outlook in the future study of both normal and abnormal growth (Chapter IX).

The last chapter of this book rounds out the work that has gone before. Chemical syntheses, metabolic reactions, differentiation and organization at the molecular and cellular levels must all be regulated so that growth may be a coordinated series of events. Our knowledge of these regulatory mechanisms stems largely from studies on the adrenal gland. This pillar in the structure which the book attempts to blueprint centers in the concluding chapter, "The Adrenal Gland, a Regulatory Factor."

This book is an outgrowth of a conference on "The Chemistry and Physiology of Growth" that was held in Princeton in September 1946 in celebration of the Bicentennial of Princeton University. Since the conference, the authors have had the opportunity to change their papers in the light of comments made then, or to comment on the work of others. Thanks are here extended to the many persons whose work made the conference and this book possible, and special thanks are due to the authors of these contributions for their generous help throughout.

ARTHUR K. PARPART

Princeton University
January 1948

CONTENTS

The Chemistry and Physiology of Growth

I. ENZYMES AND THE SYNTHESIS OF PROTEINS

BY JOHN H. NORTHROP[1]

EVIDENCE is rapidly accumulating to show that all enzymes and at least some viruses are proteins. As a result, three fundamental problems, which previously appeared unrelated, may now be considered as one general problem—the synthesis of proteins.

Solution of the problem requires the explanation of a large number of very precise experimental facts on the one hand, and is confined by strict theoretical limits on the other. Each tissue of each species of plant or animal can form its own special proteins, and these proteins are characteristic of the organ as well as of the species. It is probable, therefore, that millions of different proteins exist. In some cases, as in the production of antibodies, or adaptive enzymes, the presence of a foreign compound may cause the organism to develop a new protein having a very definite and specific relation to the compound which caused its production. In other cases the presence of a foreign protein (viruses) results in the formation of more of the foreign protein. The formation of the type-specific nucleic acid (Avery, MacLeod, and McCarty, 1944) is an example of a similar reaction.

Whatever the mechanism of protein production, there is every reason to believe it is controlled in one way or another by enzymes, and there is good experimental evidence that many enzymes behave as theoretical catalysts. The synthesis of proteins, therefore, must presumably obey the laws of catalysis, and any hypothesis purporting to account for the synthesis of proteins by means of enzyme reactions must conform to the general theory of catalysis.

The correctness of this assumption is of fundamental importance since, if true, it rules out many synthetic reactions which might otherwise be possible. The experimental evidence for the correctness of the statement is therefore reviewed in the following section.

I. THEORY OF CATALYSIS

The fundamental law of catalysis was formulated by Ostwald, who defined catalysts as substances which affect the rate but do not affect the equilibrium state of a reaction.[2]

[1] The Rockefeller Institute for Medical Research, Princeton, New Jersey.

[2] Bredig (1902) modified the definition slightly in order to make it conform more

Theoretical proof of this statement was furnished by van't Hoff, who showed by thermodynamic reasoning that a substance which is not changed during the course of a reaction cannot affect the equilibrium condition (cf. Taylor, 1925, p. 318).[3]

This peculiarity of catalytic reactions also furnishes the most useful method of determining experimentally whether a reaction is catalytic or not. If the substance under investigation acts as a catalyst (enzyme) the addition of increasing concentrations of the substance will not change the final products formed, but only the time required for the reaction to take place. If, on the other hand, the substance takes part in the reaction, the nature or quantity of the products, as well as their rate of formation, will change as the quantity of substance added is increased.

Since the catalyst does not change the equilibrium, it follows that if any one catalyst causes the reaction to proceed from *A* to *B* then any other catalyst which affects the reaction must also cause the reaction to proceed in the same direction and to the same extent. The assumption of "hydrolyzing" and "synthetic" catalysts for the same reaction under the same conditions is therefore in contradiction to theory (cf. Borsook, 1935).

It is true, however, that under some conditions the addition of a catalyst can greatly increase the concentration of one or more reaction products and it is for this reason that enzymes are so important in cellular reactions and catalysts so useful in industry. Assume the presence of a substance *A*, which decomposes "spontaneously," i.e. with the liberation of energy, to form a series of substances, *B, C, D, E*—and that at equilibrium all substances are present in equal concentration. Since they are all in equilibrium with *A* they are also in equilibrium with each other. If *A* is then allowed to decompose spontaneously the system at the end of the reaction will contain equal concentrations of *A*, *B*, *C*, *D*, and *E*, and no catalyst can cause any further change to occur. Now suppose that *A* alone is present and a catalyst is added which accelerates the

closely to the experimental facts. He defined catalysts as substances which do not appear in the reaction products in simple stoicheiometric relationship.

[3] This conclusion is strictly true only when the catalyst is present in low concentration and does not combine with any of the components of the reaction. If the catalyst is present in high concentration it may affect the activity of the components in the same way as would any indifferent substance and so change the equilibrium (Taylor, 1925). If the catalyst combines with one or more components the equilibrium will also be changed (Euler, 1925, p. 305). Enzymes in general are present in such low concentration that effects on the activity are negligible. Many enzymes combine with the products of reaction, however, and in such cases this combination may affect the equilibrium point, although the principal result is to slow down the rate of reaction so that experimentally it may be very difficult to reach final equilibrium.

reaction $A \rightleftharpoons B$ but not the other reactions. Addition of this catalyst will result in a rapid formation of B, until $[A]=[B]$ with the result that a large amount of B and very little C, D, or E is formed. The reaction is not at equilibrium and if allowed to proceed by itself will end at the same state as though the catalyst had not been added. In the meantime, however, large amounts of B have been formed due to the catalyst and could be removed. Similarly, if the concentration of B is greater than that of A at the beginning, addition of the same catalyst will now cause A to be formed until $[A]=[B]$ and very little C, D, or E will be formed. The reaction does not take place in the direction $B \rightarrow A$ simply because of the addition of the catalyst but because the energy changes require the reaction to go in this direction, and the catalyst merely accelerates the rate. In this case, also, the high concentration of A is only temporary, since if left to itself the same final equilibrium conditions, $A=B=C=D=E$, will be reached.

The condition may be roughly exemplified by the arrangement shown in Fig. 1. The five tubes are connected at the top by open tubing and at the bottom are all connected to tube A by means of tubes which can be closed. If tube A is now filled with water to position 1, the water will

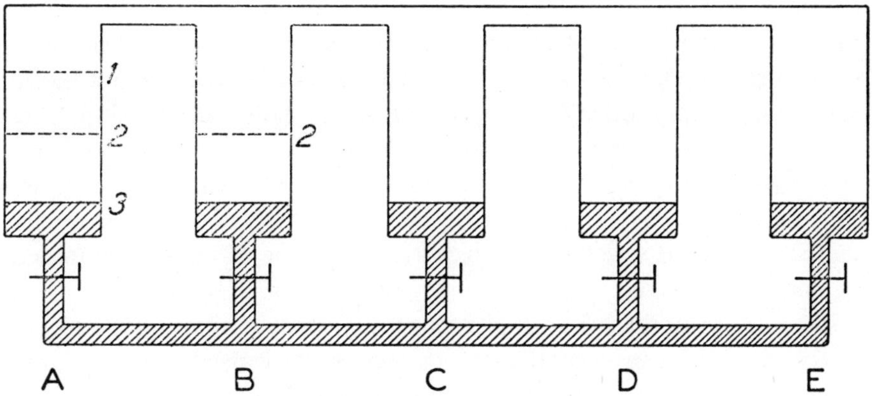

Figure 1. Enzyme reaction model.

distribute itself until it is at the same level in all tubes, since this is the equilibrium condition. If the connecting tubes at the bottom are closed at the beginning the water will distribute itself slowly through the vapor phase. This corresponds to the "spontaneous" reaction rate. If all the bottom tubes are opened the same final condition will be reached much more rapidly. This corresponds to what will happen if a catalyst is added which accelerates all the reactions. If, however, only the tube connecting *A* and *B* is opened the level of *B* will rise to position 2 and no measurable change will occur in the levels of *C*, *D*, and *E* during this time.

The preceding example resembles more or less the hydrolysis of proteins; hydrolysis with acid leads to complete hydrolysis and results in the liberation of the amino acids and, hence, any other catalyst must also lead only to amino acids *at equilibrium* under the same conditions. Pepsin or trypsin, however, increase the rate of hydrolysis of some protein linkages much more than others, so that for all practical purposes the reaction stops before any amino acids are formed, although it is still far from equilibrium.

Enzymes as Theoretical Catalysts

The agreement between enzyme reaction and catalytic theory has been much more fully confirmed in connection with carbohydrates and esters than with proteins.

The first experimental synthesis by an enzyme was reported by Croft Hill, who found that a disaccharide which he thought was maltose was formed by the action of maltase on concentrated glucose solutions. The substance, however, was eventually identified by Emmerling as iso-maltose which is formed, under the same conditions, by acid catalysis. Iso-maltose is not hydrolyzed by maltase and this fact led Armstrong to suggest that enzymes synthesize substances which they do not hydrolyze. This assumption is contrary to the law of catalysis since it predicts that the direction of the reaction, and, hence, the equilibrium condition, vary with the catalyst. It was found later that the "maltase" preparation used by Croft Hill contained emulsin which does hydrolyze iso-maltose and, hence, also synthesizes it. This is one of many examples of the confusion caused by the use of impure enzyme preparations.

Bourquelot studied the simpler case of glucoside hydrolysis and synthesis and found in several cases that enzymes caused the reaction to proceed towards the equilibrium point from either side. This furnished experimental proof of the statement that enzymes do not change the equilibrium point.

Kastle and Loevenhart showed that the hydrolysis of esters by lipase also agrees with theory.

Pottevin found that pancreatic lipase synthesizes esters from oleic acid and methyl alcohol and that the equilibrium condition was independent of the concentration of enzyme. (For review of the older literature see Euler, 1925, p. 295; and Bayliss, 1925.)

Borsook and Schott have found (1931) from a study of oxidation potentials that the enzyme, fumarase, which catalyzes the reaction, succinate=fumarate, also acts as a theoretical catalyst.

The synthesis of glycogen described by Cori and Cori (1939) is another example, since in this case the reaction proceeds in either direction, depending upon conditions. The synthetic reaction is remarkable in that a trace of the end product, glycogen, is required to start it. The enzyme, phosphorylase, however, is also necessary. The reaction as a whole therefore appears to be an example of autocatalysis superimposed on an enzyme reaction.

The experimental results with esters and glucosides in the presence of enzymes are therefore in good agreement with theory. In the case of the proteolytic enzymes, however, the results are more complicated, and there is some doubt as to the occurrence of protein synthesis in the presence of proteinases.

Various Possible Synthetic Reactions

The various methods which have been suggested to account for the synthesis of proteins may be roughly classified as follows:

1. Purely catalytic synthesis, no energy added.
   a. Equilibrium point shifted in favor of protein by change of concentration or other condition (plastein formation).
   b. Removal of protein as soon as formed. Insoluble proteins —plastein? Synthesis of anilides (Bergmann and Fruton, 1941). Protein surface films (Langmuir and Schaefer, 1938).
2. Energy added so as to shift equilibrium in favor of protein.
   a. Coupled reaction (cf. Borsook and Dubnoff, 1940, hippuric acid).
   b. Synthesis from building stones other than amino acids.

Purely Catalytic Synthesis

This mechanism (Wasteneys and Borsook, 1930) is the simplest and the most thoroughly investigated.

THEORY. Proteins are hydrolyzed by enzymes and therefore they must also be synthesized by the enzyme which caused hydrolysis, since catalysts must accelerate the reaction equally in both directions. In order actually to obtain proteins in this way, therefore, it is only necessary to find conditions under which significant quantities of protein will exist at equilibrium. Theoretically, more protein will exist in concentrated than in dilute solution. This follows from the law of mass action which states that the equilibrium constant for the reaction $A=nB$ is defined by the equation $K=[B]^n/[A]$ or $[A]=[B]^n/K$. $[A]$ and $[B]$ are concentrations at equilibrium, and n is the number of molecules of B formed from A. If $[A]$ is very small compared to $[B]$ then the concentration of A at equilibrium will increase in proportion to the nth power of the total concentration. Thus if it be assumed that a protein is hydrolyzed to 100 amino acids, increasing the total concentration 10 times should increase the concentration of unhydrolyzed protein 10^{100} times. The fact that there are undoubtedly a number of intermediate steps in the hydrolysis does not change the results, since according to thermodynamic reasoning the condition at equilibrium must be independent of the path by which the equilibrium is reached.

EFFECT OF *p*H. It is possible to assume that the equilibrium is shifted by changes in *p*H of the solution, since the energy change is small in any case and could be significantly affected by the energy of ionization of the products of the reaction.

EXPERIMENTAL RESULTS. Danilewsky found that a precipitate (plastein) was formed when a preparation of pepsin was added to a concentrated solution of the products of pepsin hydrolysis of protein. The problem was carefully investigated by Wasteneys and Borsook (1930). They confirmed the importance of the concentration, although the experimental result is not exactly as expected by theory since it is necessary also to change the *p*H. In dilute acid solution the plastein is digested, whereas the synthetic reaction had an optimum rate of *p*H 4.0. There is a decrease in amino and carboxyl groups.

The total amount of plastein formed is greater when greater amounts of enzyme are added. This is apparently contrary to theory, but it is very difficult to be sure that equilibrium is reached, especially with low concentrations of enzyme in the presence of reaction products. The products act as inhibitors and greatly decrease the reaction rate.

Wasteneys and Borsook also found that the reaction was accelerated by the presence of some emulsifying agents (benzaldehyde) and took place to some extent without the enzyme. This observation is difficult to understand from the point of view of synthesis unless these substances help remove the protein from solution. Wasteneys and Borsook consider that the protein is synthesized as a soluble protein which soon changes to an insoluble (denatured?) form and precipitates from the solution. If this is correct then the emulsifying agents may act by accelerating the denaturation reaction since many proteins are rapidly denatured at interfaces.

Plastein is also formed by the action of trypsin or papain on a solution of insulin which has been previously digested by pepsin (Haddock and Thomas, 1942).

There seems little doubt that some synthetic reactions occur during plastein formation, but whether proteins are produced is still uncertain. According to Folley (1932) no substance having a molecular weight of over 1000 is formed. This result has been confirmed by Ecker (1947).

Taylor (1907 and 1909) reported the synthesis of protamine sulfate from a completely digested solution of the compound. The enzyme preparation was a glycerine extract of liver. The substance formed was identical, or very similar, to the original protamine.

The relation of the plastein to the original protein may be tested by hydrolyzing enzyme proteins. Plastein is formed in concentrated solutions of trypsin or pepsin which have been hydrolyzed by pepsin. The plastein, however, has no enzymatic activity, nor does it have the general properties of the enzyme protein from which it is derived. In these cases, therefore, the plastein obtained is not related to the protein originally hydrolyzed (Northrop, 1947).

There is some experimental evidence that proteins may be formed by enzymatic synthesis under the proper conditions.

Removal of Protein From Reaction Mixture. The preceding section summarizes attempts which have been made to synthesize proteins by changing the concentration of the reaction mixture so as to favor the presence of the protein.

Protein could also be accumulated if the quantity in equilibrium with the hydrolysis products could be constantly removed. This operation would require energy but would result in the formation of indefinite quantities of protein.

The synthesis of anilides discovered by Bergmann and Fruton (1944) is probably an example of this type of reaction.

Robertson (1926) suggested that protein might be synthesized at liquid interfaces since the protein is concentrated there. The experiments of Langmuir (1939) and Gorter (1937) have shown that native proteins, although soluble in bulk, form an extremely insoluble surface layer at the air-water interface. If this layer were constantly removed a new one would form and in this way protein would be obtained. Langmuir has suggested (Langmuir and Schaefer, 1938) that the molecules already present on the surface could act to regulate the formation of more identical molecules. This mechanism also predicts that the synthesis of proteins would be associated with some type of structure such as is present in cells which would provide the necessary surface.

This hypothesis solves both the energy and the specificity problem and appears to be the most reasonable and simplest mechanism so far suggested. It deserves the most careful consideration. The accumulation of protein in the surface layer is an undoubted fact, so that only the specific synthesis is in doubt. Unfortunately no experimental verification has been obtained as yet for this step. It must be remembered, however, that negative results are very unconvincing in experiments of this type, since even after positive results have been reported it is sometimes necessary for the problem to be studied for years before conditions are sufficiently well established so that the results may be obtained at will.

Synthesis by the Addition of Energy

Coupled Reactions. Many cases are known in which a reaction which requires energy and hence does not occur alone will take place if another reaction which liberates energy takes place simultaneously. Such reactions control carbohydrate metabolism (Meyerhof, 1944; and Kalckar, 1944). In these reactions part of the energy liberated by hydrolysis or oxidation is used for a synthetic reaction so that, although the total change results in the liberation of energy, some steps occur which require energy. Each step is controlled by a special enzyme. Phosphoric acid esters are concerned in these energy exchanges and it is reasonable to suppose that similar esters may take part in the synthesis of proteins, as Bergmann and Fruton (1944) have suggested. Many other possible sources of energy exist, but no direct experimental evidence has been found to connect the synthetic and energy-providing reaction. The synthesis of hippuric acid from glycine and benzoic acid in the presence of liver slices is perhaps the nearest approach to such a reaction. Borsook and Dubnoff (1940) found that rapid synthesis occurred in the presence of intact liver cells. Extracts or even minced tissue failed to cause the reaction.[4] It seemed evident that the energy was obtained from respiration and hence that the reaction would stop if the respiration were stopped, and hydrolysis should then occur instead of synthesis. Addition of HCN, which poisons the respiratory enzymes, did stop synthesis, but it also stopped hydrolysis. This disconcerting result may be due to poisoning of the enzyme which catalyzes the benzoic acid reaction, al-

[4] It is perhaps significant that the three fundamental reactions, protein synthesis, photosynthesis, and fixation of nitrogen all require energy and all are so far inseparable from intact cells.

though such an effect would not be expected since only the respiratory enzymes are known to be inactivated by HCN. The result might be due to the fact that hydrolysis and synthesis are both caused by the same enzyme.

Schoenheimer's beautiful experiments (1942) have shown that proteins are continuously being synthesized and hydrolyzed so that it is quite possible that part of the energy for synthesis of proteins is obtained from the hydrolysis of other proteins or peptides.

The enzymes which hydrolyze proteins (pepsin, trypsin, papain, cathepsin, etc.) do not carry digestion very far. The hydrolysis is completed by the various peptidases. Part of the energy liberated by the hydrolysis of the peptides could, therefore, be returned to the protein $\rightleftharpoons$ peptide system and so result in the synthesis of proteins. Such a system is highly organized and hence would be difficult to isolate in good working order. It could not function indefinitely without an external source of energy.

An interesting reaction which may be an example of this mechanism has been studied by Behrens and Bergmann (1939). They found that neither acetyl-*dl*-phenylalanyl glycine nor glycyl-*l*-leucine alone are hydrolyzed by papain, but that glycine and leucine are formed when both peptides are present. Similar results have been reported by Abderhalden and Ehrenwall (1933). Behrens and Bergmann were also able to show that an intermediate compound, acetyl phenyl-alanyl glycyl glycyl leucine, is formed and then hydrolyzed. They consider that the tripeptide acts as "co-substrate" for the hydrolysis of glycyl leucine and that the energy required for the formation of the intermediate compound is obtained from the hydrolysis of glycyl leucine. The tripeptide appears to be a "co-enzyme," however, since its concentration is unchanged at the end of the reaction, rather than a "co-substrate," which is a substance changed by the reaction.

The reaction is an excellent example of a catalytic reaction in which the acceleration is due to the formation of an intermediate compound. The overall change is glycyl leucine $\rightleftharpoons$ leucine and glycine and hence the formation of the pentapeptide does not enter into the energy relation at all, since thermodynamic equilibria are entirely independent of the path by which they are reached. As Petrie (1943) has pointed out, the reaction is actually an hydrolysis and does not by itself serve as a source of the synthetic substance. Nevertheless, the reaction is of interest since it shows that the enzyme does catalyze the synthetic reaction. If a mechanism were present which would remove the pentapeptide, the reaction could be used as a source of the substance. In this case the energy for synthesis would be furnished by the mechanism which removed the peptide.

SYNTHESIS OF PROTEINS FROM COMPOUNDS OTHER THAN AMINO ACIDS. Several reactions are known (cf. Bergmann and Fruton, 1944) which could give rise to proteins. Alcock (1936) notes that plants can form proteins from sources other than amino acids. He claims with some justification that two entirely different mechanisms for protein synthesis would not be expected and suggests that animals first change the amino acids to the hypothetical building stones which are common to both plants and animals. An "ur-protein" molecule is then synthesized and the other proteins derived from this "ur-protein." This hypothesis is logically attractive but must be amplified somewhat to take into account the energy changes required (cf. page 10).

If it could be shown that the compounds taking part in these reactions are formed from amino acids by the addition of energy and that these compounds then reacted to form proteins with the liberation of energy, the mechanics of synthesis would be greatly clarified. Synthesis of proteins from such high energy compounds has been discussed by Delbrück (1941) and Gulick (1944). Such a reaction would account for the fact that N or C isotopes supplied in any amino acids are very soon found in all amino acids (except lysin) (Schoenheimer, 1942). This is difficult to understand if the proteins are synthesized from amino acids, but would be expected if amino acids are first changed to other intermediates before synthesis occurs.

At present, however, there does not appear to be any experimental evidence for the existence of such compounds, or reactions, or the necessary catalysts in biological material. The enzymatic synthesis of glutamic acid from NH_3 and the keto acid (Euler, Adler, Günther, and Das, 1938) is probably the nearest approach to such a reaction so far discovered. Glutamic acid can be converted into several other amino acids by transamination (Braunstein and Kritzmann, 1937; Braunstein, 1939). The transmigration of the methyl groups in certain amino acids, studied by DuVigneaud and others (reviewed in Borsook and Dubnoff, 1943) furnishes another partial mechanism for equilibria between amino acids. These reactions can account in part for transformation of one amino acid into another but do not supply a complete synthetic reaction for proteins.

Formation of Native as Distinct from Denatured Proteins

It was stated earlier in this discussion that the reaction, protein and water $\rightleftharpoons$ amino acids is nearly complete and very little protein remains at equilibrium. This is experimentally true in the case of denatured proteins but merely an assumption as far as native proteins are concerned. Most native proteins are digested very slowly by proteases and it is possible to assume, as Linderstrom-Lang and Jacobsen (1941) have suggested, that the first step in the hydrolysis is the formation of denatured protein from the native protein. Linderstrom-Lang attempted to answer the question by a study of the temperature coefficient of hydrolysis of native and denatured protein, but the results were inconclusive. The hydrolysis of a protein like trypsin or chymotrypsin, which exists in an equilibrium between native and denatured forms (Northrop, 1939), should furnish evidence in this connection.

Haurowitz and collaborators (1945) have reported that globular proteins, such as egg albumin and serum globulin, are not hydrolyzed by trypsin unless they are denatured, whereas "fibrous" proteins, such as fibrin and myosin, are hydrolyzed at about the same rate in either native or denatured forms.

If enzymatic hydrolysis does require preliminary denaturation, then it could be assumed that *native* proteins can be synthesized without adding energy, i.e. by a purely catalytic reaction. Energy is required to denature the protein which then hydrolyzes with the liberation of energy. The entire cycle would require three reactions:

1. Amino acids $\overset{\text{enzyme}}{\rightleftharpoons}$ native protein + energy.
2. Native protein + energy $\rightleftharpoons$ denatured protein.
3. Denatured protein $\overset{\text{enzyme}}{\rightleftharpoons}$ amino acids + energy.

It is also possible to assume that *denatured* protein is formed first from the amino acids and that the native proteins are then formed from this denatured protein. This cycle would be as follows:

Amino acids + energy $\rightleftharpoons$ denatured protein
Denatured $\rightleftharpoons$ native protein + energy

The direction in which the reactions proceed is determined by the energy changes and the rate of the reaction by the enzymes present.

Specificity of Synthesis

The preceding outline shows that several mechanisms are known which could result in the synthesis of proteins without coming in conflict with any of the accepted theories of chemical reactions. Even if some of these overall mechanisms are correct, there still remains the problem of the regulation of the reaction so that the desired protein only is obtained and not simply a random assortment of proteins.

Nothing is to be gained, in the absence of experimental evidence, by assuming a series of enzymes, each controlling the synthesis of a single protein, since we are then faced with the mechanism of synthesis of the enzymes themselves. These are presumably also proteins and we are therefore back at the beginning again.

The known proteases hydrolyze practically all (denatured?) proteins and hence must catalyze the synthesis of all proteins, so that these enzymes can hardly account for the formation of any one protein. For instance, horse hemoglobin and pig hemoglobin are both presumably hydrolyzed completely to amino acids by the autolytic enzymes of either the horse or pig. If, therefore, conditions were found in which the reaction were reversed, addition of either of these enzymes would result in the formation of at least two hemoglobins, horse and pig. In the animal, however, this does not occur. Horse hemoglobin only is formed in the horse and pig hemoglobin only is formed in the pig.

It is possible that the specificity of protein synthesis is not due to the specific action of the enzyme but simply to the chemical nature of the protein itself. Most chemical reactions are specific in that, as a rule, very few and often only one of a large number of possible reaction products actually appear. Thus if the reaction between NaOH and HCl is treated purely statistically, there are a great many possible reaction products. Actually, however, only two products are formed—NaCl and H_2O. This "specificity" is not due to any catalyst but simply to the fact that the products, NaCl and H_2O, are the only stable ones under the condition of the reaction. It is therefore possible that the formation or synthesis of specific proteins is regulated by the difference in stability of the many structurally different proteins rather than by specific catalysts.

Autocatalysis

Autocatalysis is the only known chemical reaction by which indefinite quantities of a required substance may be obtained without assuming the existence of several other substances, the origin of which must also be accounted for. Autocatalysis, therefore, offers the only escape from the dilemma of the preformationists and has been invoked by a number of workers. Troland (1917) developed a general theory of cell formation on this basis. He concluded that complicated molecules reproduced themselves by a process similar to crystallization. The molecular structure of the underlying layer determined the structure of the superimposed layer. Troland also reviewed earlier and less definite suggestions of autocatalysis. Koltzoff (1928) developed a similar theory. He concluded that antibody formation was a special example of such a reaction. Koltzoff also realized the fact that an autocatalytic reaction cannot start itself, but must have at least one molecule of the reaction product present at the beginning (cf. Northrop, 1937). He assumed the "genonema" contains at least one molecule of each protein of the species. It is difficult to avoid the necessity for this assumption but it may be that, if the total number of different kinds of protein molecule present in the individual were known, genonema and even the sperm and ovum would be too small to contain them. Gulick (1938) suggests that protein molecules are biscuit-shaped discs and that this structure is impressed on each succeeding molecule somewhat as a coin is stamped by a die. He points out, however, that the contact must be "front to back", as otherwise the impression of the disc and not the replica would be obtained. Langmuir and Schaefer (1938) have also suggested the "template" method of regulating synthesis. Pauling's (1940) theory of antibody formation is somewhat similar. Darlington (1944) ascribes the formation of "plasma germ" and viruses to a similar autocatalytic reaction.

Stanley (1938a) suggests that a protein-nucleic acid complex is the simplest structure capable of autocatalytic synthesis. He ascribes the specific synthesis to a crystallization-like reaction similar to that suggested by Troland.

A Working Hypothesis for the Synthesis of Proteins

The preceding section has summarized briefly theories and experimental results relating to protein synthesis. These theories account more or less satisfactorily for the specificity of the reaction but (with the exception of the system of Langmuir and Schaefer) fail to provide for the necessary energy. Since there is every reason to believe that energy is required, the synthesis cannot be accounted for by a simple autocatalytic reaction, nor can it be said that any of the various types of proteins or viruses "synthesize themselves," unless it be further assumed that they are formed from unknown, high energy compounds. They may be able to direct the synthesis so that the proper protein is formed, but the energy must come from some other reaction, so that at least two reactions, and not a single autocatalytic reaction, are necessary.

The working hypothesis outlined in the rest of this discussion is an attempt to formulate the simplest assumption that is adequate to account for the known facts. It is quite probable that future experiments will force modification to be made and may render the entire mechanism untenable. If the hypothesis leads to the discovery of new facts it will have served its purpose even though the new facts destroy the assumptions which led to their discovery.

In general the hypothesis may be stated as follows:

The formation of proteins occurs in two steps. The first step consists in the synthesis of one or more "type" proteins ("proteinogens") ("ur-protein" of Alcock) which are specific for the species and perhaps for the organ. This step requires energy which may be obtained from a coupled reaction or from preliminary formation of high energy building stones (not amino acids). It is "autocatalytic" in that the structure of this proteinogen is determined by itself. No experimental evidence for this step exists at present. It probably takes place in the cell, since respiration and similar reactions that liberate energy occur usually in the cell.

Since even the existence of such a protein is purely hypothetical, discussion of its properties is hardly warranted. The only essential properties are that stable proteins may be formed from it without the addition of energy and that its own formation is autocatalytic. There may be several such proteinogen structures, one for each general group of proteins. They probably contain nucleic acids. The virus proteins may be special examples of these proteinogens.

Spiegelman and Kamen (1946) have suggested that nucleoproteins are the controlling factor in protein synthesis, and present evidence indicates that the energy may be supplied by the phosphoric acid of the nucleoprotein. Muller (1945) has made a similar suggestion.*

The possibility that the proteinogen has the characteristics of a denatured, rather than a native protein, was suggested by M. Kunitz. This assumption would account for the observed immunological reactions as well as the energy requirements (cf. pp. 10, 27), since denatured proteins do not cross-react with the native proteins, nor do they show

* Recent experiments by Reiner and Spiegelman (1948) and by Price (1948) bear out this hypothesis and also indicate a relationship between virus and adaptive enzyme formation. Spiegelman and Reiner found that a desoxyribonucleic acid fraction from yeast cells stimulates the formation of adaptive enzyme in yeast. Price finds that the same fraction greatly increases the yield of virus particles per cell, formed by *S. muscae*.

such marked specificity. Therefore cross-reactions between the proteinogen and the native proteins derived from it would not be expected.

It is probably unnecessary to assume that the molecular weight of the proteinogen is equal to or greater than that of the largest molecules which are to be formed from it since Svedberg (1937) and others have found that very large protein molecules like the hemocyanins dissociate very easily and reversibly. The formation of these very large molecules from molecules the size of normal proteins, therefore, requires little or no energy (Northrop, 1938, p. 363).

It is also unnecessary to assume that any appreciable quantity of this protein exists at any one time. It could be decomposed as rapidly as formed and still act as an intermediate between the energy-requiring step and the specific step in protein synthesis.

In the second step the individual proteins are formed by a catalytic or an autocatalytic reaction from this proteinogen. This reaction does not require energy and may occur anywhere.

The formation of an enzyme from its precursor is an example of such a reaction. It is the only mechanism so far discovered whereby proteins may actually be produced *in vitro*. The reactions are specific and are adequate to account for the formation of proteins in general, provided the precursor is present. The exact chemical changes involved in these reactions are not known.

The proposed mechanism accounts for the group specificity of proteins from the same organ and species. Assumption of one (or a few) synthetic reactions is simpler than assumption of a separate synthetic reaction for each individual protein. It has the further advantage that it is in keeping with the gradual development of the cell, since, once the proteinogen is formed, other proteins can readily be derived from it (cf. Troland, 1917). If each protein must be synthesized individually, each one requires an energy source and the evolution of a series of related proteins is more difficult.

Madden and Whipple (1940) have pointed out that partial synthesis of all proteins in one organ (the liver?) would greatly simplify the mechanism of tissue metabolism since it would then be unnecessary for every cell of every tissue to contain the complete protein-synthesizing system. According to the present mechanism, the proteinogen could well be synthesized in one organ and distributed from there to the various other organs and tissues where it would be modified into the specific proteins needed.

It may be noted in this connection that many autocatalytic and many coupled reactions are known, but no autocatalytic coupled reactions have been described, so far as the writer is aware. If every cell in every tissue had such coupled reactions occurring for each protein, the chances of the discovery of the reaction would be far greater than if it occurred only in a special organ or organs.

In the following sections an attempt is made to describe the formation of normal proteins, viruses, adaptive enzymes, and antibodies in terms of this working hypothesis.

II. Formation of Normal Proteins

It is becoming increasingly evident that some sort of equilibrium exists between the various amino acids and the various proteins or both (Madden and Whipple, 1940; Schoenheimer, 1942). Schoenheimer demonstrated by means of isotopes that both the N and C atoms of an ingested amino acid appear in nearly all the proteins and amino acids of the body in the course of a few days. This result is totally unexpected from the viewpoint of the classical theory of protein synthesis and indicates the existence of a dynamic equilibrium (cf. Borsook and Keighley, 1935), since such equilibria would result in a rapid distribution of the atoms. It is probable that, as Schoenheimer has suggested, some unknown building stones take part in the equilibria, as otherwise it is difficult to account for the transfer of the C atoms from one amino acid to another.

Whipple and Madden's results are extremely interesting in this connection. These workers find that dogs may be kept in health when the only source of nitrogen is plasma injected into the vein. The animal must, therefore, be able to form any required protein from the protein present in plasma without preliminary hydrolysis by the enzymes of the digestive tract. Whipple and Madden consider the results to show that an equilibrium exists between plasma protein and the cell proteins. They suggest that large polypeptides may be the intermediate compounds in the equilibrium.

A similar equilibrium between the serum proteins themselves has often been suggested, and there is considerable evidence to show that these proteins can be transformed into each other.

Pederson has found by ultracentrifugal studies that a lipoid-containing protein of enormous molecular weight, which he calls X-protein, exists in serum. He assumes that this protein is in equilibrium with the albumin and globulin fractions. It is true that albumin and globulin are not in equilibrium with each other when isolated from serum, but this fact by no means proves that they are not in equilibrium with each other in the circulating blood.

Moore, Shen, and Alexander (1945) have found a special protein in chicken embryos which is not found in adult chicken serum.

The various equilibria indicated by these experimental results may be described in terms of the present hypothesis as follows:

Amino Acids
⥯ (Energy added somewhere between amino acids and proteinogen.)
(Intermediate?)
⥯ (Autocatalytic reaction.)
Proteinogen
± Enzyme ⥯ ⥯ (Autocatalytic or catalytic reaction.)
Protein A, Protein B, Protein C, etc.

Since all the proteins are in equilibrium with the proteinogen they are in equilibrium with each other.

The quantity of proteinogen present at any time may be very small and hence difficult to detect.

The proteinogen is probably synthesized only in cells. The formation of the usual proteins from proteinogen may occur anywhere.

No experimental evidence for the formation of normal proteins in this way is at hand. It is possible that such a reaction would be observed if an organism could be completely freed from one of its normal proteins. The re-addition of this protein to the organism should then result in renewed production.

The formation of enzymes from their precursors is exactly analogous to the hypothetical reaction assumed for the formation of all normal proteins, and hence one of interest in this connection. The precursors of the enzymes must themselves be formed from the proteinogen molecules, and an indication of such a reaction has been noted by Allen, Ray, and Bodine (1938).

Formation of Enzymes from Their Precursors

Trypsin from Trypsinogen. This reaction agrees quantitatively with the theory for a simple autocatalytic reaction. The velocity is affected by acidity and temperature in the same way as is the formation of chymo-trypsin from chymotrypsinogen (Kunitz and Northrop, 1935) or the hydrolysis of other proteins by trypsin. At *p*H 5.0 and 8°C. the value of the constant is 14.6, i.e. with unit concentration of trypsinogen the active trypsin would increase 14 times per hour. As in the case of chymo-trypsin there is no evidence for the splitting off of any part of the molecule during this activation reaction, nor is there any marked change in chemical composition so that the change from the inactive protein to the enzyme is probably due to some internal rearrangement in the structure of the molecule. However, the work on this question is still in the preliminary stage. The autocatalytic nature of this reaction was correctly described by Vernon but subsequently denied by Bayliss, Starling, and others.

Kunitz (1938) has recently found that trypsin is also formed from trypsinogen at *p*H 5.0 by a proteolytic enzyme secreted by a mold (penicillium). Trypsin is only slightly active at this *p*H, so that the activation curve is no longer autocatalytic but simply logarithmic as is the activation of chymo-trypsinogen. Trypsin obtained in this way is identical with that formed by autocatalytic activation. The mold enzyme, therefore, must attack the trypsinogen molecule at the same place as does trypsin.

The formation of trypsin from trypsinogen is also accelerated by enterokinase. The reaction trypsinogen → trypsin is therefore catalyzed by three different catalysts—trypsin itself, mold kinase, and enterokinase.

Activation of Trypsin by Enterokinase. Pancreatic juice as secreted is inactive and becomes active *in vivo* only when mixed with the contents of the small intestine. The subject has been controversial for many years. Kunitz (1939b) has reinvestigated the problem with purified materials and has found that the conflicting results are due to the fact that an inert protein, as well as the active enzyme, are formed simultaneously during the reaction. The relative amounts of the two substances obtained depend on the acidity of the solution. In slightly acid solution nearly all the precursor is transformed into the active enzyme and very little inert protein is formed. In neutral or slightly alkaline solution considerable amounts of inert protein are formed.

The entire reaction may be quantitatively predicted on the assumption that the rate of formation of the active enzyme is proportional to the concentration of precursor and of activator, while the rate of formation of the inert protein is proportional to the concentration of trypsin and of the precursor. The relative rates of the two reactions are determined by the acidity.

Pepsin from Pepsinogen (Herriott, 1938). The reaction at *p*H 4.65

is autocatalytic and hence is caused by pepsin itself. About 15 per cent of the nitrogen is split off during this reaction. So far as is known, pepsin attacks only peptide linkages, so there is reason to believe that the rupture of one or more peptide links in the precursor leads to the formation of the active enzyme. If swine pepsinogen is activated by chicken pepsin, swine pepsin is formed (Herriott, Bartz, and Northrop, 1938). The structure responsible for the species specificity of the enzyme is therefore present in the precursor.

Relation of Protein to Precursor. The exact changes in chemical structure which occur during these reactions is not yet known. It is known, however, that the protein formed may be quite distinct in chemical, physical, and immunological properties from its precursor.

It is probable that all amino acids present in the enzyme must also be present in the precursor although, in view of the remarkable reactions discovered by Schoenheimer, even this conclusion may be questioned.

Immunological Relationship of Chymo-trypsin, Chymo-trypsinogen, and Beef and Pig Trypsin (Ten Broeck, 1934). The tests were carried out by the Dale technique for anaphylactic reactions. Chymo-trypsinogen was found to be distinct from chymo-trypsin. Guinea pigs sensitized with chymo-trypsinogen were negative in most cases when tested with chymo-trypsin, although four animals out of twelve gave cross-reactions. No cross-reactions were observed when the pigs were sensitized against chymo-trypsin and tested with chymo-trypsinogen. There is a slow formation of chymo-trypsin from chymo-trypsinogen even in the absence of trypsin (Kunitz and Northrop, 1935) and it is quite possible that enough chymo-trypsin was formed in this way during the experiment to sensitize some animals, and so produce the cross-reaction.

Trypsin from beef pancreas was distinct from trypsin from pig pancreas and both were distinct from chymo-trypsin and chymo-trypsinogen.

Pepsinogen and Pepsin (Seastone and Herriott, 1937). These tests were carried out by the precipitin reaction. They are complicated by the fact that pepsin is inactivated rapidly at the *p*H of the blood so that the antibody formed is presumably caused by the denatured rather than the native protein. It is also possible that traces of pepsin are formed from pepsinogen after injection into the blood stream.

Pepsinogen antisera reacted with pepsinogen but did not react with pepsin nor with serum protein from the homologous species.

Antipepsin sera reacted with pepsin and also to a small extent with pepsinogen. They do not react with serum proteins of the homologous species.

These results show that there is at best only a faint cross-reaction between the enzyme and its precursor, even when the experiments were carried out with concentrated solutions of the two pure proteins. Had the tests been carried out with crude tissue extracts of the precursors it is probable that no cross-reaction between the enzyme and its precursor would have been observed, owing to the small quantities present.

Specificity of the Reactions. Each enzyme, as a rule, has its own precursor, but Bodine (1945) has reported that different tyrosinases may be formed from protyrosinase by different methods. The enzyme precursors isolated so far, therefore, do not represent the hypothetical proteinogen, but must be considered a separate step between this and the active enzyme. The formation of the precursor of tyrosinase has been studied by Allen, Ray, and Bodine (1938) and found to be an autocatalytic reaction. The enzymes whose precursors have been isolated are secretory enzymes and are required in high concentration at special times. It is possible that relatively large quantities of the precursors of these enzymes are to be found in the organs for this reason, since they serve as a convenient source of supply from which the active enzymes can be obtained rapidly at any time. The cellular enzymes and the normal proteins, on the other hand, are not required suddenly and there is no necessity for a rapidly available reserve. This difference in the requirements may account for the fact that the precursors for the secretory enzymes only have been found in quantity.

III. Formation of Viruses

There is good reason to believe that some viruses, at any rate, are proteins (for a review of this work see Stanley, 1940).

No experimental evidence exists, so far as the writer is aware, to distinguish the formation of virus from the formation of normal proteins. Both are found only in or on living cells and virus production is accompanied usually, if not always, by protein production. Experimentally the only difference between the two processes is that "normal" proteins are always produced by the cells but that the production of some virus protein may be started by introduction of the protein from outside the organism (Northrop, 1937; Darlington, 1944). Phage produced by lysogenic strains of bacteria and "indigenous virus" of plants and animals are always produced just as are "normal" proteins. It can, of course, be assumed (cf. Gratia, 1938) that infection occurred sometime in the past, but there is no direct evidence for this assumption. No means is known to free lysogenic strains from their viruses.

It would be expected, then, that if an organism could be completely freed from a protein, it would no longer produce that protein until reinoculated with it. The suggestion was made (Northrop, 1938) that the substance producing type-specificity in pneumococci was an example of such a reaction. The analogy between virus and the substance regulating type-specificity in pneumococci was recognized by Gratia (1936). Since then the substance has actually been isolated (Avery, MacLeod, and McCarty, 1944). It turns out that it is a desoxyribonucleic acid instead of a protein, but the mechanism of formation may well be the same.

Virus proteins possess the following characteristics which must be accounted for by any hypothesis for their formation:

1. They increase only in the presence of appropriate living cells.
2. Production of virus is usually associated with growth and active metabolism of the host cells (Bordet and Jaumain, 1921; Krueger and Northrop, 1930; Zinsser, 1937; Gratia, 1938; Howe and Mellors, 1945, this paper contains other relevant references).

This correlation between metabolism and virus production is strong evidence that the synthetic part of the reaction is carried out by the cell in conjunction with the synthesis of normal proteins, since the viruses themselves do not contain the enzymes necessary to carry out these reactions.

If it should be found that virus production can take place *only* during cell division, the analogy between virus and gene suggested by many workers (cf. Beadle, 1945) would be greatly strengthened, since genes also multiply only during cell division, whereas structural proteins are formed continuously.

All hypotheses concerning the production of virus proteins (other than those which consider the virus to be a cell) assume some sort of autocatalytic reaction. Some writers (Troland, 1917; Stanley, 1938b; Caspersson, 1939; Jordan, 1944) assume that the virus can synthesize itself from simple building stones. This mechanism requires energy and entails a series of assumptions as to the source of energy and the manner in which it is made available (cf. page 10).

Bordet (1931) pointed out that a simpler and equally adequate hypothesis is that the virus is formed from a precursor. This assumption does not involve any energy mechanism in addition to that already present in the host cells and has the further advantage that there is a close experimental analogy in the formation of enzymes (Gratia, 1922; Northrop, 1938). Darlington (1944) has recently suggested a similar mechanism. Potter and Albaum (1943) consider viruses to be "misplaced enzymes."

Viruses have also been considered as obligate parasites. This point of view is ably presented by Rivers (1939) and by Burnet (1945), who consider that a virus is a fragment of a degenerate bacterial cell. This fragment has maintained or developed the power of multiplying in the presence of the host cell. Actually the original source of the virus is the only significant difference between these points of view and that of Bordet, who considered that the virus was originally part of a host cell, rather than of a parasitic bacterial cell. It is quite possible that both ideas are correct since the definition of virus is at present largely (and quite arbitrarily) based on size. Vaccinia, for instance, is similar to bacteria in many ways, and would not be considered a protein even though its peculiar physiological and pathological properties were not known.

On the other hand the well-defined crystalline viruses, tobacco mosaic and especially bushy stunt of tomatoes, are typical proteins in many respects and would never be considered as related to parasites were it not for the pathological changes which they cause.

Final decision between the possibilities probably cannot be made until the reaction can be produced *in vitro* and in the absence of host cells (cf. Seifriz, 1939).

3. Viruses are formed as a rule only in the presence of cells from related species, but may sometimes increase in the presence of cells of widely separated species.
4. Viruses are immunologically distinct from the host proteins.
5. The virus protein produced in the presence of appropriate living cells is usually identical with that used for inoculation but may be different.
6. Inoculation with two viruses results in production of only one, as a rule. In the case of bacteria (Delbrück and Luria) the virus which is inoculated first is produced and not the second.
7. No evidence of metabolism has been found.
8. Viruses are of high molecular weight and contain nucleic acid. Neither of these characteristics, however, suffices to distinguish them from normal proteins.

9. Some viruses at least (bacteriophage, Northrop, 1939; tobacco mosaic, Loring, 1940) can form saturated solutions. This result shows that the virus can exist in solution or, more precisely, that the liquid containing the virus is a single phase.

Such solubility experiments afford a decisive test of the question whether the virus molecules are in "solution" or not. From the kinetic point of view any particle is a molecule and there is no distinction experimentally or theoretically between particles and molecules (cf. Taylor, 1925, p. 1279). For this reason the word molecule seems appropriate until evidence is found contradicting its use.

From the point of view of thermodynamic equilibria as stated in Gibb's phase rule, however, there is a sharp and definite distinction between suspensions and solutions. A suspension is a two-phase system and a solution is a single-phase system. It is perfectly possible to decide, therefore, whether a virus "solution" consists of one or two phases.[5]

Quantitative measurements have not been made on other viruses, but any virus which crystallizes must exist as a saturated solution, in equilibrium with the crystals.[6]

Such solubility experiments have not been carried out with larger viruses, such as vaccinia, but it appears very unlikely that any indication of a saturated solution would be obtained in these cases.

[5] There is some uncertainty as to the application of the phase rule to colloidal solutions, but the results obtained with proteins (Northrop, 1939) show clearly that the rule may be applied to these compounds. The following experimental results are possible in a study of the solubility of viruses:

1. The concentration of virus in the liquid is zero or variable and no equilibrium can be shown to exist between the solid and liquid. Such systems are suspensions, but no further conclusion can be drawn since the phase rule cannot be applied to systems which are not in equilibrium.

2. The concentration of virus in the liquid is constant and in equilibrium with the solid virus.

a. The concentration of virus in the solution is independent of the quantity of solid virus present. The system consists of two phases and two components, and the virus therefore is a single component. The virus is in solution in the liquid phase. No such results have been reported.

b. The concentration of virus in solution varies with the amount of solid. There are two phases and three or more components. The virus is in solution in the liquid phase. The virus preparation consists of two or more components.

This is the result obtained with bacteriophage. (Northrop, 1938) and tobacco mosaic preparations (Loring, 1940), although the bacteriophage curve is as good as that obtained with many normal proteins.

[6] The fact that the solutions may be concentrated in the centrifuge does not affect this conclusion, since this simply amounts to increasing the force of gravity and any solution may theoretically be concentrated in this way. The gravitational field simply represents another degree of freedom.

The properties of viruses outlined above may be accounted for in terms of the present hypothesis, as follows:

Specificity of Virus and Host. Viruses in general increase only in related host species. It is known from immunological studies that the proteins of related species are similar and some may be identical.

It would be expected, therefore, that the virus precursor (proteinogen) would be similar or identical in related species but different in widely separated species and hence that a virus would usually be produced only in related species. This is usually the case.

The anatomical and physiological differences between species increases as development proceeds, and embryos of different species are much more alike than adults. It is quite possible that this similarity of the embryos extends to the structure of the proteins and that embryo proteins are more alike than adult proteins (cf. Pedersen, 1945). This may account for the fact discovered by Murphy (1912) that heterologous transplants will grow in chick embryos. In general such transplants do not grow when made in adults.

The fact that chemical differentiation of the proteins proceeds more or less parallel to the morphological differentiation of the organs is clearly shown by the results of Burke *et al.* (1944). These workers followed the development of organ antigens in growing embryos and found that organ specificity was closely related to the growth of the organs themselves.

Woodruff and Goodpasture (1931) found that many viruses would multiply in chick embryos, although they would not do so in the adult chicken. These observations suggest that the proteinogens of embryos are more closely related than are those of the adult animal and hence are able to serve as precursors for heterologous proteins while the proteinogens of the adult animal cannot.

Moore, Shen, and Alexander (1945) have found that plasma from chicken embryos contains principally one high molecular weight protein which is absent, or present in low concentration, in adult plasma. Pedersen (1945) has found a characteristic protein (fetuin) to be present in foetal calf plasma, and he has suggested that this protein may be a precursor of the normal serum proteins.

Burnet, Freeman, Jackson, and Lush (1941) and Dale have objected to the assumption of a precursor on the grounds that the specific character of the virus would then be determined by the host and not by the virus, as is usually considered to be the case.

The character of a virus which has been modified chemically *is* determined by the host. Thus tobacco mosaic virus with oxidized SH groups (Anson and Stanley, 1941) or the acetyl or phenyluriedo derivative of tobacco mosaic (Miller and Stanley, 1941) all give rise to the original tobacco mosaic protein and not to the modified form which was used for inoculation.

In these cases the type of virus formed is determined by the host. The changes in yellow fever virus after passage through mouse brain (Lloyd, Theiler, and Ricci, 1936) and of equine encephalitis virus grown in pigeons (Traub and Ten Broeck, 1935) may be other examples in which the nature of the virus is determined by the host. This subject has been reviewed in detail by Stanley (1943). Stanley considers these changes in the virus to be mutations.

On the other hand, proof of the identity of the virus produced by two different hosts rests in most cases only on immunological and pathological properties. The same tests fail to differentiate between pepsin formed from cattle or swine pepsinogen (Seastone and Herriott, 1937). The differences between the two enzymes were established only by a mixed solubility determination (Northrop, 1939, p. 32). Such tests have not been made with virus proteins. In any case there is nothing to indicate that the same precursor may not exist in many different species. For example the crystalline lens proteins of many species are indistinguishable by immunological reactions and may be identical in various species.

The assumption of a precursor has also been criticized on the grounds that its presence would be detected by immunological tests. This objection sounds very reasonable on *a priori* grounds, but as a matter of fact only a questionable cross-reaction could be found between pepsin and its precursor, pepsinogen[7] (Seastone and Herriott, 1937), or chymo-trypsin and its precursor, even with the classical Dale technique (Ten Broeck, 1934; cf. page 16). These experiments were carried out with concentrated solutions of both enzyme and precursor and it is probable that no cross-reaction would have occurred with crude tissue extracts. If the proteinogen is a denatured protein, such cross-reactions would not be expected.

Recent work (Knight, 1946) has demonstrated that influenza virus grown on chicken embryos reacts with normal chicken protein antiserum,

[7] These results, however, are open to the objection that the pepsin was denatured at the *p*H of the blood which the pepsinogen was not. Denaturation is known to modify the specificity of proteins, and hence the failure to obtain a cross-reaction in this case is not conclusive. The failure to differentiate bovine and swine pepsin is also complicated by the fact that both proteins were denatured. This objection, however, does not apply to Ten Broeck's results with chymo-trypsin.

whereas influenza virus grown on mouse lung does not. The mouse-lung virus, on the other hand, reacts with normal mouse antiserum but not with normal chicken antiserum. After a single passage through chicken embryos the mouse virus reacts with normal chicken protein antiserum.

The results of Gye and Purdy with Fujinama tumor virus from different hosts are very similar to those found by Knight for influenza virus. In both cases the results agree with the predictions based on the analogy between the formation of viruses and normal proteins.

It is not necessary to assume that any appreciable amount of precursor is present at any one time since it would serve its purpose as a link between the synthetic and specific part of the reaction even though it were only present in traces as an intermediate product.

Luria, Delbrück, and Anderson (1943) were unable to find any large molecules, comparable to those of the virus, in uninfected cells. They consider that this result proves that no precursor exists. There is no necessity, however, for the precursor to be of the same molecular weight as the virus since association of proteins to larger molecules is known to occur readily. This step is quite different from the synthesis of the primary bonds in the smaller units (Northrop, 1938).

Adaptation of Viruses.* In certain cases inoculation of a different host results in the production of a different virus after a longer or shorter time (equine encephalitis, yellow fever). This result may also be accounted for on the basis of a precursor. Assume that Host 1 contains Precursor P_1 and that Host 2 contains a slightly different precursor, P_2. The original virus forms more of itself by reaction with Precursor P_1 but forms more or less of a slightly different virus, V_2, as well as more of itself, by reaction with Precursor P_2. The reaction in Host 1 is $P_1 + V_1 \rightarrow V_1$. In Host 2 three reactions occur, $P_2 + V_1 \rightarrow V_1$; $P_2 + V_1 \rightarrow V_2$; also $P_2 + V_2 \rightarrow V_2$. If the production of V_1 and V_2 proceeds at exactly the same rate, both will be formed indefinitely. It is extremely improbable that two such reactions have exactly the same rates. If the rates differ, that reaction with the higher rate will eventu-

* Sudden changes in the properties of a virus are usually referred to as "mutations" of the virus. As far as the writer is aware, there is no evidence to indicate whether the mutation occurs in the virus or in the host cell. The experiments show that a different virus appears as a result of some reaction in the host-virus system, but whether this is due to a mutation in the host cell which then produces a different virus, or a mutation in the virus itself, cannot be determined. If the mutation occurred first in the virus, it would be expected that a chemically modified virus would reproduce itself, but as discussed above, it does not. Attempts to cause mutation by irradiating tobacco mosaic virus were also unsuccessful, whereas mutations did arise if infected leaves were irradiated (Kausche and Stube, 1939).

ally be the only one which takes place. If $P_2 \rightarrow V_2$, therefore, is more rapid than $P_2 \rightarrow V_1$, the virus finally produced will be practically pure V_2, the length of time required depending on the relative reaction rates.

An example of such a transformation is furnished by Gratia's interesting results (1936) with B. *megatherium*. This organism exists in two forms, "lysogenic" and "sensitive." The lysogenic strain constantly produces a virus *T*, which does not cause lysis of the parent lysogenic strain but does cause lysis of the sensitive strain. Subculture of plaques of this virus on the sensitive strain gives rise to two different viruses, the original *T* and a new one, *C*, which causes lysis of both strains. Similar results have been reported by Luria (1945).

It does not appear possible to make any generalization on theoretical grounds as to whether or not the products of an "autocatalytic" reaction are identical with the original catalyst. The hydrolysis of esters by acids is an autocatalytic reaction and may give rise either to more of the same acid or to different acids, depending on conditions. Thus the hydrolysis of ethyl acetate, propyl acetate, and butyl acetate by means of acetic acid will result in the formation of more acetic acid; but the hydrolysis of ethyl butyrate and ethyl propionate by acetic acid will result in the production of butyric and propionic acids. Although hydrolysis is caused primarily in this case by the hydrogen ion, which is the same in all acids, the same relation holds presumably for proteins; the ability to produce certain reactions is primarily a property of certain groups and these groups may be common to many different proteins.

Quantitative Data on the Formation of Bacterial Viruses (Bacteriophages). If viruses are formed from precursors in the cell it follows that the quantity of virus formed in any one cell should be constant, provided no increase in the precursor occurs during the reaction. Viruses in general, and bacterial viruses in particular, increase most rapidly in the presence of growing host cells; and under these conditions the number of infected host cells is constantly changing and more precursor is presumably being formed so that the amount of virus formed per host cell cannot be determined. Krueger and his collaborators (cf. Krueger and Scribner, 1939) found that bacteria viruses could be produced under certain conditions without cell growth, and Krueger and Baldwin (1937) obtained occasional virus increase in cell-free solution. Krueger concluded that the virus was produced from a precursor in the bacterial cell. Spizizen (1943) reports virus production in dilute solutions of glycine anhydride without cell growth. Delbrück and Luria separated virus production from cell growth by dilution of the culture after inoculation with virus. All of these workers agree that the bacterial cell must be in good physiological condition and also that each cell forms a certain definite amount of virus.

Delbrück and Luria (1942) have investigated the reaction in a series of extremely careful and interesting papers and have obtained a detailed picture of the steps in the increase of virus. Their results may be summarized as follows:

1. Virus is rapidly absorbed by the host cell, the rate of absorption being proportional to the concentration (cf. Krueger, 1931).

2. No increase in infective centers, as indicated by plaque count, occurs for a definite time interval which is characteristic of the virus.

3. The number of infective centers then increases rapidly for a few minutes until a certain definite number have appeared for each host cell.

4. The greater the number of virus molecules originally present in the bacterial cell, the more virus molecules are finally formed, but increasing the original number causes much less than the proportional increase in the number formed. Delbrück and Luria assume that the increase in virus observed with increasing infection is due to secondary complications and that actually the number of virus molecules formed per host cell is independent of the infecting number.

5. Two different viruses, if grown separately, produce the same number of molecules from the same bacterial cell.

6. If the bacteria are inoculated with two different viruses simultaneously, only one virus increases. If the bacteria are inoculated with two viruses successively, only the one which is first inoculated increases.

Delbrück and Luria conclude that the "intracellular" increase in virus is limited by the availability of some substrate, and that liberation takes place when the reaction has run to completion. They assume that a "key enzyme" exists which controls the reaction and which can react with only one virus molecule. This peculiarity accounts for the fact that either virus but not both may increase, and also for the fact that the time for the growth cycle is nearly independent of the number of infective units originally present. It does not account for the fact that the number of molecules formed increases as the number of infective units increases.

These results may be explained equally well by the precursor mechanism, without assuming the existince of a key enzyme. The following assumptions are necessary:

1. Each host cell contains a definite quantity of precursor. The quantity varies depending on the physiological state of the cell.

2. Virus is formed from this precursor in accordance with the usual

autocatalytic equation $dv/dt = KV(V_e - V)$ when $V_e =$ total concentration at end of reaction and $V =$ concentration[8] of virus at time t.

The explanation of the next step depends on whether or not the variation in the number of virus molecules formed varies with the number of molecules with which the bacteria is infected. The direct experimental results show that it does, but Delbrück and Luria suggest that this is due to secondary reactions.

If it be assumed that the number of molecules produced does vary with the size of the infection, then it must be assumed that the time required for liberation of the virus depends to some extent on the bacteria cell and not entirely on the number of virus molecules present. There is some indication of this, since changing the temperature changes this time interval in the same way as it does the growth rate of the bacteria.

We may assume that virus is liberated after a definite time has elapsed and that this time is determined both by the state of the bacteria cell and by the nature of the virus, but not by the number of molecules of virus. The theory of autocatalytic reaction applied to these conditions predicts that the number of molecules produced in the interval will increase more slowly than does the original number. Fig. 2 shows a series

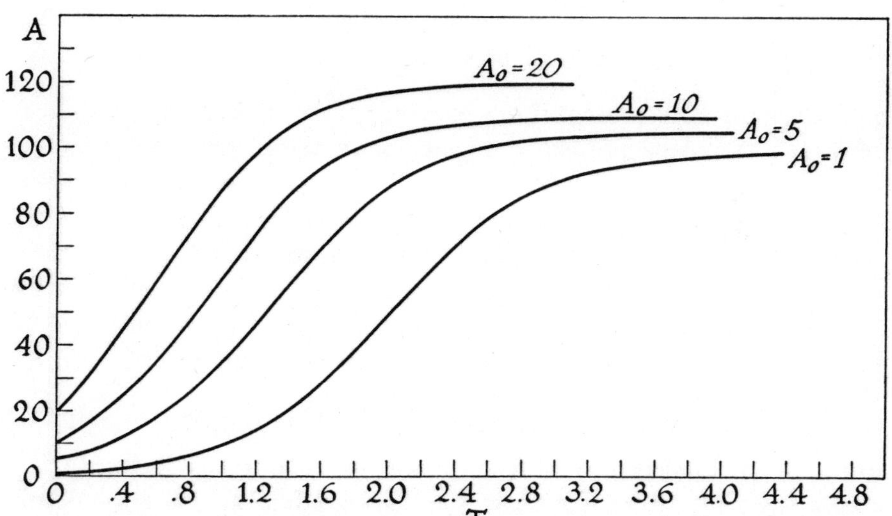

Figure 2. Effect of initial concentration of the end product on an autocatalytic reaction.

[8] It may be recalled in this connection that most enzyme reactions deviate from simple theory in that the velocity does not change in the expected way with either the enzyme or substrate concentration. In the case of the autocatalytic formation of pepsin from pepsinogen, for instance, increasing the concentration has much less effect on the velocity than predicted by the equation given above (Herriott, 1938).

of such autocatalytic curves calculated by the autocatalytic equation for the conditions that 1, 5, 10, or 20 per cent of the final maximum number of molecules are present at the beginning. The curves show that after the first part of the reaction the increase in the number of molecules formed after a given time interval increases much more slowly than does the original number. For instance (Table I), after 2.0 minutes, 50

TABLE I

Value of A corresponding to different values of A_o after different time intervals

Calculated from Autocatalytic Equation, $T = \frac{2.3}{KA_e} \text{Log} \frac{A(A_e - A_o)}{A_o(A_e - A)}$

$K = .023$; $A_e = 100 + A_o$; $A_T =$ Number of particles present at time T (mins.)

A_0	$A_{1.5}$	$A_{2.0}$	$A_{2.5}$	$A_{3.0}$
1	25	50	75	90
5	65	90	100	103
10	90	104	107	108
20	110	115	117	118

molecules would be present if one were present at the beginning, whereas only 115 would be present if twenty were present at the beginning. After 3.0 minutes the corresponding figures are 90 and 118. These figures are similar to those actually found by Delbrück and Luria. They were calculated on the assumption that the reaction followed the theoretical course. Actually the effect of changing the total concentration on the rate of enzyme reactions is usually less than that predicted by simple theory. This effect would tend to decrease the difference in the figures given above.

3. If the number of molecules liberated per bacterial cell is actually constant and independent of the infective quantity as Delbrück and Luria suggest, then it is necessary to assume that the liberation occurs as soon as the virus-forming reaction is completed. This assumption predicts also that lysis occurs as soon as the ratio of virus to bacteria reaches a definite value. This result agrees with the experiments of Krueger and Northrop (1930). If liberation of virus occurs as soon as this constant amount of virus is formed, the time required should be shorter as the number of infective units increases, although the varia-

tion is small. Interpolating from Fig. 2 again, it is found that if the reaction is started with 1 per cent of the final quantity, 3.4 minutes are required to produce 95 per cent, whereas if it is started with 5 per cent about 2.5 minutes are required, while 10 per cent requires 1.8 minutes. This variation is greater than that actually found by Delbrück and Luria for the total elapsed time, but the variation in the time of formation itself may be greater than is indicated by the variation in observed time. In other words, the observed time is the time required for the virus to be formed plus the time required for it to be liberated from all the infected cells; $T_{\text{obs.}} = T_{\text{form.}} + T_{\text{lib.}}$. If an appreciable part of this time is required for liberation to occur, then relatively large changes in formation time would cause only slight variations in the total observed time. Information in this connection could be obtained by titrating the phage by the dynamic method (Krueger, 1930), which determines the actual quantity present and not simply the number of infective centers.

4. The results obtained with a mixture of viruses may be explained in a similar way. The first virus added increases so that by the time the second virus is added, more or less of the precursor is used up. The rate of formation of the second strain will then be in proportion to its concentration (assuming equal rates of reactions), so that if 50 units of virus *A* had been formed at the time one unit of *B* was added, approximately 50 times as much *A* as *B* would be formed subsequently.

This mechanism differs from that of Delbrück and Luria principally in that it predicts that a cell can produce both strains simultaneously, whereas Delbrück and Luria's hypothesis predicts an all or none result.

The data actually show that both viruses are produced by the culture, but this is explained by Delbrück and Luria as due to production of the different viruses by different cells. Anderson (1942) found that chick embryo cells may contain inclusion bodies of two viruses. Sylverton and Berry (1936) have reported similar results. It is quite certain that some cells, therefore, may produce two viruses simultaneously.

5. In experiments with partially inactivated virus Luria and Delbrück (1942) found that exposure to ultraviolet light changes the virus so that it no longer produces lysis, but it still stops bacteria growth and prevents formation of the active form of the same or of another virus.

The inactivated virus does not increase. This result may be due simply to the fact that the bacteria do not grow in the presence of the inactivated virus, since anything which stops bacterial growth generally stops virus formation.

It is possible to assume, however, that the partially inactivated virus reacts with the precursor and forms more "inactive virus," but this virus is not liberated from the cell since no lysis occurs. From this point of view the interference caused by the "inactivated" virus is similar to the interference between two viruses and is due to the destruction of the precursor.

IV. Formation of Adaptive Enzymes

The formation of normal proteins and enzymes and of viruses outlined in the preceding section is purely autocatalytic and results in the formation of more of the same molecules. The formation of adaptive enzymes or antibodies, however, is not purely autocatalytic, since the molecules formed are different from those originally present. The hypothesis, therefore, must be modified to account for the formation of these compounds.

Enzymes are sometimes formed as a specific response to the presence of the corresponding substrate or its decomposition products. Enzymes formed in this way are called "adaptive enzymes" (Karström, 1930).

The production of galactase by yeast is the best known example but many similar cases are known (cf. Yudkin, 1938; Knight, 1936; Dubos, 1939). "Normal" yeast cannot ferment galactose, but Dienert (1900) found that yeast acquires this power if grown in the presence of galactose. It was thought at first that this result was due to selection of yeast cells during growth of the culture. Later work established the fact that the enzyme can be formed without any increase in the number of cells. If this observation is correct the fermentation cannot be due to selection of yeast cells, but must be due to production[9] of a new enzyme.[10] The formation of the enzyme is always associated with the formation of protoplasm, even though no cell division occurs (Dubos, 1939).

In some cases, like the polysaccharide-splitting enzyme of Dubos (1939), the enzyme displays extreme specificity, in fact the specificity of this enzyme is more precise than the immunological tests.

Formation of the enzymes may be caused by the decomposition prod-

[9] There does not appear to be any advantage in distinguishing between "production" and "activation" of an enzyme. An inactive enzyme, strictly speaking, is not an enzyme at all.

[10] Sevag (1946) considers that no new enzymes are actually formed but that the new substrate increases the activity of the normal enzyme in some unexplained way. According to this point of view the phenomenon is simply a change in the specificity of the reaction similar to that which occurs when various amino acids are added to peptidase systems (cf. page 11).

ucts of the substrate. The production of yeast invertase is brought about by glucose as well as by sucrose (Euler and Cramer, 1913).

The production of the enzyme is not inherited but stops when the substrate is removed.

The potential ability to produce the enzyme, however, is inherited as a mendelian character (Lindegren, Spiegelman, and Lindegren, 1944).

Yudkin suggests that the new enzyme is always present in minute amounts in all cells and is in equilibrium with its precursor. Addition of the substrate or products of the reaction or other compounds which react with the enzyme will decrease the enzyme concentration. This disturbs the equilibrium and hence more enzyme will be formed. This explanation is adequate and also accounts for the fact that substances which combine with the enzyme, other than the substrate, may bring about its production.

It includes the assumption that minute amounts of all possible "adaptive" enzymes exist in normal cells and also that the equilibrium is such that very little enzyme is present. It is further necessary to assume that equilibrium is reached slowly, since the new enzyme does not appear at once. The hypothesis predicts that the new enzyme should be obtainable *in vitro* if the precursor were available.

Actually there is no experimental evidence of an equilibrium between an enzyme and its precursor, since the known reactions of this type run to completion (*in vitro*) as far as can be determined. It is possible to assume equilibrium, but it must be very far in the direction of the active enzyme.

If the reaction were autocatalytic the combination of the enzyme with the products of reaction would decrease the rate of reaction and hence delay the attainment of equilibrium conditions. The formation of trypsin from trypsinogen, for instance, is stopped by the addition of trypsin inhibitor (Kunitz and Northrop, 1936).

The assumption that the substrate acts by changing the course of formation of the normal (closely related) enzyme appears to be equally probable. This assumption has the advantage that closely analogous cases are known and also that it may be applied to the formation of antibodies, a similar reaction. It has the disadvantage that it does not explain the specificity of the new enzyme quite so definitely.

According to this hypothesis the normal enzymes are continuously being formed from their precursors in the cell by an autocatalytic reaction. When a new substrate or other compound which combines with the enzyme is added, it may act as a co-enzyme and change the course of the reaction so that a slightly different enzyme is formed as well as the usual enzyme. When the new substrate is no longer present the reaction returns to its normal course and the normal enzyme alone is formed. Examples of similar reactions in which the course of an enzyme reaction is affected by the presence of substances other than enzyme and substrate are well known. Thus either trypsin or an inert protein may be formed during the autocatalytic formation of trypsin from trypsinogen; the nature of the product depends on the *p*H of the solution (Kunitz, 1939a). Different tyrosinases may be formed from protyrosinase under different conditions (Bodine, 1945). Abderhalden and Ehrenwall (1933) found that addition of certain peptides caused glycyl-leucine to be hydrolyzed by trypsin, whereas the reaction did not occur when trypsin and glycyl-leucine alone were present. The experiments were extended by Bergmann and his collaborators (cf. page 11). The added peptides appear to act as co-enzymes.

The specificity of the amino acid oxidases is completely changed by the addition of other proteins or amino acids to the reaction mixture (Edlbacher, 1946).

The products of an enzyme reaction usually combine with the enzyme so that the fact that a reaction occurs between the enzyme produced and the substance which caused its production does not differ qualitatively from many enzyme reactions. The distinguishing factor in the present case is that one of the products of the reaction is itself an enzyme.

V. Formation of Antibodies

Antibodies are considered to be modified serum globulins which are changed in such a way as to react with the homologous antigen (cf. Marrack, 1938; Zinsser, Enders, and Fothergill, 1939; Burnet, Freeman, Jackson, and Lush, 1941; Landsteiner, 1945; and Sevag, 1945). There is no doubt that antibodies are proteins, but their relationship to the serum globulins appears to be somewhat uncertain owing to the fact that neither antibodies nor serum globulins have been obtained in pure form (cf. Roche, Derrien, and Mandel, 1944a and 1944b).[11] Any

[11] Crystalline diphtheria antitoxin (Northrop, 1941) is probably a pure protein and is not a globulin, although it cannot be separated from the globulin fraction if mixed with it by any known procedure except precipitation with the homologous antigen. This result would prove that antibodies are distinct chemically and immunologically from the normal serum proteins, were it not for the fact that the toxin-antitoxin complex used to prepare the crystalline antibody was digested with trypsin. It is possible to assume, therefore, that the pure crystalline antibody does not represent the natural antibody but is an hydrolysis product.

Antibodies are reported to be associated with different serum proteins in different

hypothesis for the formation of antibodies must account for the following facts (Burnet, Freeman, Jackson, and Lush, 1941; Landsteiner, 1945):

1. They are proteins which react specifically with their antigens.

2. They may be produced by so many different antigens that it is extremely unlikely that even one molecule of all possible antibodies exists in the normal animal.

3. They are produced in great quantity as compared to the quantity of antigen which causes their formation.

4. Antibody continues to be formed after the antigen has disappeared. This statement unfortunately cannot be considered an experimental fact since the evidence, although strong, is not conclusive (cf. Burnet, Freeman, Jackson, and Lush, 1941; Landsteiner, 1945). There is no doubt that antibody continues to be formed long after any antigen can be detected experimentally in the circulation, but it is possible to assume that minute quantities of antigen are still present in the cells. As Burnet points out, this question is of fundamental importance to the theories of antibody formation proposed by Ostromuislenskii (1916), Koltzoff (1928), Mudd (1932), Pauling (1940), Pauling and Campbell (1942), and Sevag (1945), since these hypotheses all include the assumption that antibody is formed as a result of some type of reaction between antigen and globulin. Harris and Ehrich state that the decomposition products of antigens occur autocatalytically.

5. The chemical nature of antibodies may be different in different species or in animals of different ages.

This statement is open to question since no antibodies have been isolated in pure form.

6. Antibodies are formed only in the animal receiving the antigen. Injection of another animal with antisera (passive immunization) results in rapid disappearance of antibody and not in the formation of more antibody. This fact precludes any purely autocatalytic mechanism for the formation of antibodies, unless secondary hypotheses are added.

7. Antibodies are formed in cells, probably in the lymphocytes (Dougherty, Chase, and White, 1944, 1945; Dougherty and White, 1945; Harris and Ehrich, 1946).

---

animals and have been found in all serum fractions (cf. Landsteiner, 1945, p. 133). These results also indicate that the antibody itself may be a distinct protein rather than a modified globulin. On the other hand, young animals which do not form antibodies readily are said to possess very small amounts of serum globulin (Orcutt and Howe, 1922).

A number of reports of the formation of antibodies *in vitro* have been made (Ostromuislenskii, 1916; Pauling, 1940) but the identity of the substance prepared in this way with the natural antibody has not been established. It must be remembered in connection with these experiments that owing to experimental difficulties *whole blood* cannot be used *in vitro*. Failure of the experiment when carried out with serum, plasma, defibrinated blood, or in the presence of anti-coagulants, therefore, does not prove that the reaction cannot occur in the circulation.

BURNET'S HYPOTHESIS. Burnet, Freeman, Jackson, and Lush (1941) consider the formation of antibodies to be analogous to the formation of adaptive enzymes. Indeed, there are many points of similarity. Antibodies and enzymes may both be divided into two general classes—normal or physiological, and acquired or adaptive. The "normal" enzymes or antibodies are present in all individuals of the species and are inherited (cf. Landsteiner, 1945, p. 132). The ability to form adaptive enzymes also appears to be inherited (Lindegren, Spiegelman, and Lindegren, 1944). "Acquired" antibodies or adaptive enzymes are formed in response to the presence of a foreign substance, and their formation stops sooner or later after removal of this substance.

Both antibodies and adaptive enzymes have a specific relation to the substance which caused their formation. In some cases adaptive enzyme reactions are more specific than the antibody reactions (Dubos, 1939).

The reaction of enzymes with substances formed during the reaction is quantitatively similar to the reaction between antigen and antibody (Northrop, 1922; and Zinsser, 1923). The antigen-antibody complex is less dissociated than most enzyme-product compounds.

According to Burnet, normal globulin is synthesized by a special proteinase. Injection of antigen results in a reaction between this proteinase and the antigen, which destroys the antigen and modifies the proteinase. This modified proteinase causes the formation of antibody instead of normal globulins, and also causes replicas of itself to be formed. The modified proteinase resulting from reaction with the antigen decreases gradually when antigen is no longer present, as do adaptive enzymes.

The qualitative and quantitative differences in antibody formation are therefore accounted for. The hypothesis also accounts for the fact that small quantities of antigens can cause the production of very large quantities of antibody and predicts that antibody formation may continue for an indefinite length of time after the antigen and even its decomposition products have disappeared.

It appears to the writer that this hypothesis predicts that injection of an animal with antiserum should lead to the production of more antibodies, since the production of the antibody-forming proteinase is assumed to be autocatalytic. These proteinases are assumed to be present in the blood and so should cause formation of antibody when blood containing them is injected into a different animal. Actually the injection of antiserum does not lead to the production of more antibodies, but on the contrary the antibody injected disappears quite rapidly. It is possible to avoid this discrepancy by means of secondary hypotheses, but it appears to be a serious objection to any purely autocatalytic mechanism. There is no experimental evidence for the existence of the "proteinases," as far as the writer is aware. In the absence of such evidence the assumption of their existence seems unnecessary and leads to the further complication of accounting for their own formation.

SEVAG'S HYPOTHESIS. Sevag (1945) has recently reviewed the various hypotheses for the formation of antibodies and has carefully and thoroughly discussed the relation of antigens to enzymes. There is no doubt that many points of similarity exist between the action of enzymes on their substrates and the formation of antibodies after injection of antigens. Sevag concludes that all proteins are enzymes and that antibodies are formed as a result of the action of the antigens (enzymes) on the normal globulins. According to this hypothesis antibodies should be readily formed by the action of antigens on normal serum *in vitro*, but this is not actually the case. This discrepancy, however, cannot be considered conclusive owing to the difficulty of carrying out such an experiment under the conditions which exist in the blood vessels.

Sevag's hypothesis also predicts that antibodies can be produced only so long as antigen is present.

Antibody Formation from Proteinogen

The writer agrees with Burnet that antibodies and adaptive enzymes are closely related and are probably formed by the same mechanism. The hypothesis describing the formation of adaptive enzymes outlined above is simpler than that of Burnet and agrees as well with the known facts of antibody formation.

The reaction may be summarized as follows:

1. The normal serum proteins are formed from the "proteinogen" by an autocatalytic reaction.

2. Antigen (or its decomposition products) acts as a co-enzyme[12] in this reaction and causes the formation of a slightly different protein, in the same way that the substrate causes the formation of the adaptive enzymes (page 34). The resulting protein reacts with the co-enzyme (antigen) which led to its formation since products of enzyme reactions in general react with the enzyme which produced them.

3. The antibody is formed almost entirely in the cells where proteinogen is synthesized, since the quantity of proteinogen present in the blood at any time is probably small.

These assumptions account for the following facts:

1. Antibodies are related to normal serum proteins and react with the antigen which led to their formation.

2. Small quantities of antigen can produce indefinite amounts of antibody.

3. Antibody production may continue after all antigen has disappeared from the circulation but not after all decomposition products of the antigen have disappeared (Harris and Ehrich, 1946). In this respect it differs from Burnet's and Sevag's hypotheses.

4. The antibody may be related to one or several normal proteins, depending on which autocatalytic reaction is affected by the antigen.

5. More antibodies will not be formed when antiserum is injected into an animal, since antibody formation itself is not assumed to be autocatalytic. This prediction also differs from Burnet's hypothesis.

6. Antibodies are not formed in any appreciable amount in the absence of cells. The proteinogen is formed in the cells, since it is only in the cells that the necessary mechanism exists for the energy supply required for synthesis. It will be transformed into the various special proteins of the circulation as soon as it comes in contact with them. The formation of antibodies will therefore occur principally in or near the synthesizing cells. All the various special proteins, including antibodies, are assumed to be in equilibrium with the proteinogen and hence are in equilibrium with each other. Therefore it should be possible to prepare antibodies *in vitro* by adding antigens to blood. The quantity of proteinogen actually present is assumed to be very small, so that the rate of reaction may be slow. There is the further technical difficulty that whole blood cannot be used (cf. page 39). Production of antibodies in this way has been frequently reported (cf., for instance, Ostromuislenskii, 1916; Pauling, 1940).

The hypothesis does not account in any simple way for the observa-

[12] Sevag (1945) assumes that the antigens themselves act as enzymes.

tion that antibody formation is sometimes accompanied by an increase in normal serum globulins.

Rate of Appearance of Antibodies in the Circulation

Injection of antigen gives rise after a few days to a rapid increase of antibody in the circulation. In some cases a second injection of antigen gives rise to a second and larger increase in antibody, especially if a small amount of antigen were injected the first time. The curves are logarithmic in form and Burnet considers this to be evidence for the existence of some process of biological multiplication. Actually the

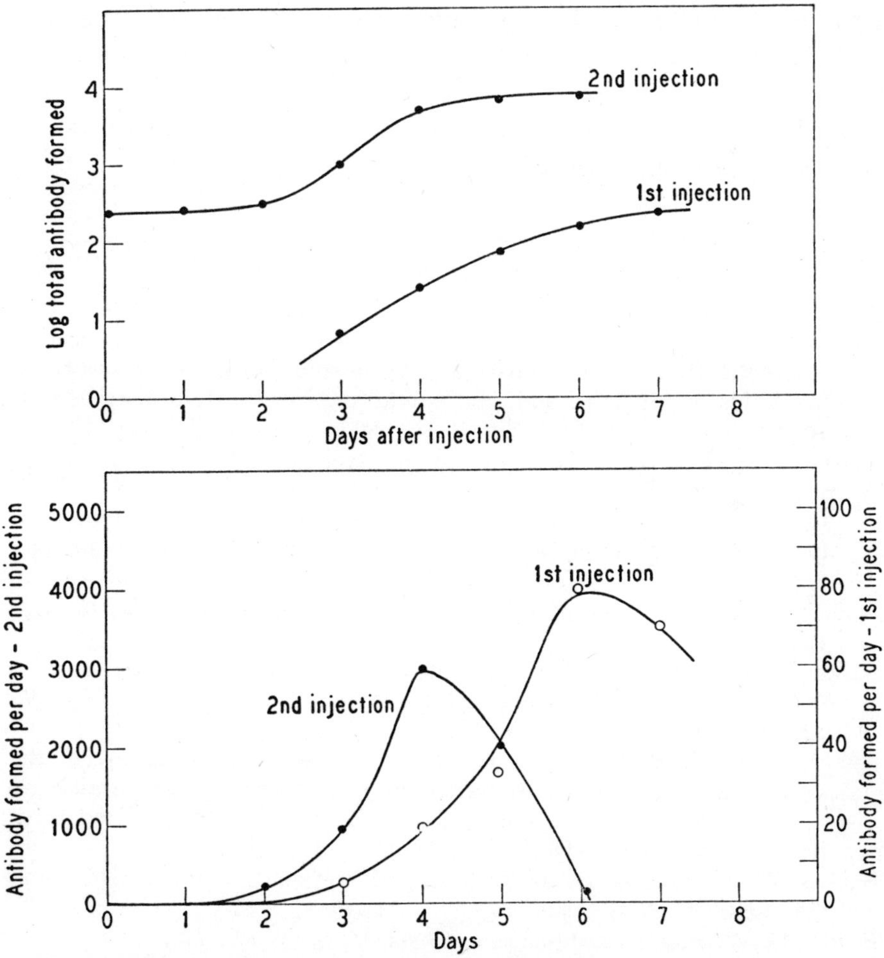

Figure 3. Production of antibody after varying lengths of time.

curves have long "lag periods" and are not strictly logarithmic except for a small part of the reaction. They agree just as well or better with the integral of the usual probability curves. (For the mathematical relation between logarithmic and probability curves, see Yule, 1910.) The case is very similar to the rate of death curves of bacterial cultures or of insects (Loeb and Northrop, 1917). The data may be made to fit a log curve if the "lag period" is neglected. The appearance of antibody in the blood, plotted as log curves (Burnet, Freeman, Jackson, and Lush, 1941), and as probability curves, is shown in Fig. 3. From the point of view of probability curves, the results indicate that the cells in which the antibody is produced die or are ruptured when antibody is released into the blood stream. The rapid and large rise sometimes noted on a second injection of antigen may thus be related to the phenomenon of anaphylaxis. The cells are sensitized by the first small injection and injured by the second. This mechanism agrees with the process of antibody formation suggested by Sabin (1939).

REFERENCES

Abderhalden, E., and E. v. Ehrenwall. Continued studies on the provocation of ereptic action in "erepsin-free" trypsin solutions prepared according to Waldschmidt-Leitz and Harteneck. IV. *Fermentforschung, 14,* 1, 1933.

Alcock, R. S. The synthesis of proteins in vivo. *Physiol. Rev., 16,* 1, 1936.

Allen, T. H., O. M. Ray, and J. H. Bodine. Enzymes in ontogenesis of Orthoptera. VI. Autocatalytic nature of formation in vivo of protyrosinase [in grasshopper eggs]. *Proc. Soc. Exp. Biol. N.Y., 39,* 549, 1938.

Anderson, K. Dual virus infection of single cells. *Amer. J. Path., 18,* 577, 1942.

Anson, M. L., and W. M. Stanley. Some effects of iodine and other reagents on the structure and activity of tobacco mosaic virus. *J. Gen. Physiol., 24,* 679, 1941.

Avery, O. T., C. M. MacLeod, and M. McCarty. Chemical nature of the substance inducing transformation of pneumococcal types. Induction of transformation by a desoxyribonucleic acid fraction isolated from pneumococcus Type III. *J. Exp. Med., 79,* 137, 1944.

Bayliss, W. M. *The Nature of Enzymes.* 5th edition, Longmans, Green and Co., London, 1925.

Beadle, G. W. Biochemical genetics. *Chem. Rev., 37,* 15, 1945.

Behrens, O. K., and M. Bergmann. Cosubstrates in proteolysis. *J. Biol. Chem., 129,* 587, 1939.

Bergmann, M., and J. S. Fruton. The specificity of proteinases. *Advances in Enzymology, 1*, 63, 1941.

Bergmann, M., and J. S. Fruton. The significance of coupled reactions for the enzymatic hydrolysis and synthesis of proteins. *Ann. N.Y. Acad. Sc., 45*, 409, 1944.

Bodine, J. H. Tyrosinase and phenols. Action of diversely activated tyrosinase on monohydric and *o*-dihydric phenols. *Proc. Soc. Exp. Biol. N.Y., 58*, 205, 1945.

Bordet, J. Theories of the bacteriophage. *Proc. Roy. Soc. London,* B., *107*, 398, 1931.

Bordet, J., A. Gratia, and D. Jaumain. Dualité du principe lytique du colibacille et du staphylocoque. *C. R. Soc. Biol. Paris, 85*, 1095, 1921.

Borsook, H. Reversible and reversed enzymic reactions. *Ergebn. Enzymforsch., 4*, 1, 1935.

Borsook, H., and J. W. Dubnoff. Biological synthesis of hippuric acid in vitro. *J. Biol. Chem., 132*, 307, 1940.

Borsook, H., and J. W. Dubnoff. The metabolism of proteins and amino acids. *Ann. Rev. Biochem., 12*, 183, 1943.

Borsook, H., and G. L. Keighley. The "continuing" metabolism of nitrogen in animals. *Proc. Roy. Soc. London,* B., *118*, 488, 1935.

Borsook, H., and H. F. Schott. The role of the enzyme in the succinate-enzyme-fumarate equilibrium. *J. Biol. Chem., 92*, 535, 1931.

Braunstein, A. E. Die enzymatische Umaminierung der Aminosäuren und ihre physiologische Bedeutung. *Enzymologia, 7*, 25, 1939.

Braunstein, A. E., and M. G. Kritzmann. Über den Ab- und Aufbau von Aminosäuren durch Umaminierung. *Enzymologia, 2*, 129, 1937.

Bredig, G. Die Elemente der chemischen Kinetik, mit besonderer Berücksichtigung der Katalyse und der Fermentwirkung. *Ergebn. Physiol., 1*, 134, 1902.

Burke, V., N. P. Sullivan, H. Petersen, and R. Weed. Ontogenetic change in antigenic specificity of the organs of the chick. *J. Infect. Dis., 74*, 225, 1944.

Burnet, F. M. *Virus as Organism.* Harvard University Press, Cambridge, 1945.

Burnet, F. M., M. Freeman, A. V. Jackson, and D. Lush. *Production of Antibodies,* Macmillan and Co., Ltd., Melbourne, 1941.

Caspersson, T. The role of nucleic acids in the cell. *Proc. 7th Intern. Genetical Congr.*, Edinburgh, Aug. 23-30, *1941*, 85, 1939.

Cori, G. T., and C. F. Cori. The activating effect of glycogen on the enzymic synthesis of glycogen from glucose-1-phosphate. *J. Biol. Chem., 131*, 397, 1939.

Darlington, C. D. Heredity, development and infection. *Nature, 154*, 164, 1944.

Delbrück, M. A theory of autocatalytic synthesis of polypeptides and its application to the problem of chromosome reproduction. *Cold Spring Harbor Symposia, 9*, 122, 1941.

Delbrück, M., and S. E. Luria. Interference between bacterial viruses. I. Interference between two bacterial viruses acting upon the same host, and the mechanism of virus growth. *Arch. Biochem., 1,* 111, 1942.

Dienert, F. Sur la fermentation du galactose et sur l'accoutumance des levures à se sucre. *Ann. Inst. Past., 14,* 139, 1900.

Dougherty, T. F., J. H. Chase, and A. White. The demonstration of antibodies in lymphocytes. *Proc. Soc. Exp. Biol. N.Y., 57,* 295, 1944.

Dougherty, T. F., J. H. Chase, and A. White. Relationship of antibody content of normal and malignant lymphocytes. *Proc. Soc. Exp. Biol. N.Y., 59,* 172, 1945.

Dougherty, T. F., and A. White. Functional alterations in lymphoid tissue induced by adrenal cortical secretion. *Am. J. Anat., 77,* 81, 1945.

Dubos, R. J. Enzymic analysis of the antigenic structure of pneumococci. *Ergebn. Enzymforsch., 8,* 135, 1939.

Ecker, P. G. Ultracentrifuge study of Plastein. *J. Gen. Physiol., 30,* 399, 1947.

Edlbacher, S. The problem of totality in biochemistry. *Experientia, 2,* 7, 1946.

Euler, H. v. *Chemie der Enzyme,* I. Verlag von J. F. Bergmann, 3rd edition, München, 1925.

Euler, H. v. *Chemie der Enzyme,* II. Verlag von J. F. Bergmann, 2nd and 3rd editions, München, 1927.

Euler, H. v., E. Adler, G. Günther, and M. B. Das. Components of dehydrogenase systems. XXI. Enzymic decomposition and synthesis of glutamic acid in animal tissues. *Z. physiol. Chem., 254,* 61, 1938.

Euler, H. v., and H. Cramer. Untersuchungen über die chemische Zusammensetzung und Bildung der Enzyme. IX. Zur Kenntnis der Invertasebildung. *Z. phys. Chem., 88,* 430, 1913.

Folley, S. J. Nature of plastein. *Biochem. J., 26,* 99, 1932.

Gorter, E. Protein-Films. *Tr. Faraday Soc., 33,* 1125, 1937.

Gratia, A. On the so-called bacteriophage. *Brit. Med. J., 296,* 1922.

Gratia, A. Dissociation du bactériophage du *Bacillus megatherium* lysogène 899 en deux variétés distinctes. *C. R. Soc. Biol. Paris, 123,* 1018, 1936.

Gratia, A. *Nature des ultravirus, Levaditi et Lépine,* Librairie Maloine, Paris, 109, 1938.

Gulick, A. What are the genes? II. The physico-chemical picture; conclusions. *Quart. Rev. Biol., 13,* 140, 1938.

Gulick, A. The chemical formulation of gene structure and gene action. *Advances in Enzymology, 4,* 1, 1944.

Haddock, J. N., and L. E. Thomas. The synthesis of plasteins by the action of trypsin and papain on digests of insulin. *J. Biol. Chem., 144,* 691, 1942.

Harris, T. N., and W. E. Ehrich. The fate of injected particulate antigens in relation to the formation of antibodies. *J. Exp. Med., 84,* 157, 1946.

Haurowitz, F., M. Tunca, P. Schwerin, and V. Göksu. The action of trypsin on native and denatured proteins. *J. Biol. Chem., 157,* 621, 1945.

Herriott, R. M. Kinetics of the formation of pepsin from swine pepsinogen and identification of an intermediate compound. *J. Gen. Physiol., 22*, 65, 1938.

Herriott, R. M., Q. R. Bartz, and J. H. Northrop. Transformation of swine pepsinogen into swine pepsin by chicken pepsin. *J. Gen. Physiol., 21*, 575, 1938.

Howe, H. A., and R. C. Mellors. Cytochrome oxidase in normal and regenerating neurons. *J. Exp. Med., 81*, 489, 1945.

Jordan, P. Zum Problem der Eiweisz-Autokatalysen. *Naturwissensch., 32*, 20, 1944.

Kalckar, H. M. The function of phosphate in enzymatic syntheses. *Ann. N.Y. Acad. Sc., 45*, 395, 1944.

Karström, H. Enzyme production in bacteria, *Lab. Butterexportges. Valio m.b. H.*, Helsinki, 1930.

Kausche and Stube. *Naturwis., 27*, 501, 1939.

Knight, B. C. J. G. *Bacterial Nutrition.* Sp. Rep. Ser. Med. Res. Coun., London, No. 210, 1936.

Knight, C. A. Sedimentable component of allantoic fluid and its relationship to influenza viruses. *J. Exp. Med., 80*, 83, 1944.

Knight, C. A. Precipitin reactions of highly purified influenza viruses and related materials. *J. Exp. Med., 83*, 281, 1946.

Koltzoff, N. K. Physikalisch-chemische Grundlage der Morphologie. *Biol. Zentr., 48*, 345, 1928.

Krueger, A. P. A method for the quantitative determination of bacteriophage. *J. Gen. Physiol., 13*, 557, 1930.

Krueger, A. P. The sorption of bacteriophage by living and dead susceptible bacteria. I. Equilibrium conditions. *J. Gen. Physiol., 14*, 493, 1931.

Krueger, A. P., and D. M. Baldwin. Production of phage in the absence of bacterial cells. *Proc. Soc. Exp. Biol. N.Y., 37*, 393, 1937.

Krueger, A. P., and J. H. Northrop. The kinetics of the bacterium-bacteriophage reaction. *J. Gen. Physiol., 14*, 223, 1930.

Krueger, A. P., and E. J. Scribner. Serial production of phage from intracellular phage precursor. *Proc. Soc. Exp. Biol. N.Y., 40*, 51, 1939.

Kunitz, M. Formation of trypsin from trypsinogen by an enzyme produced by a mold of the genus penicillium. *J. Gen. Physiol., 21*, 601, 1938.

Kunitz, M. Effect of the formation of inert protein on the kinetics of the autocatalytic formation of trypsin from trypsinogen. *J. Gen. Physiol., 22*, 293, 1939a.

Kunitz, M. Formation of trypsin from crystalline trypsinogen by means of enterokinase. *J. Gen. Physiol., 22*, 429, 1939b.

Kunitz, M., and J. H. Northrop. Crystalline chymo-trypsin and chymotrypsinogen. I. Isolation, crystallization, and general properties of a new proteolytic enzyme and its precursor. *J. Gen. Physiol., 18*, 433, 1935.

Kunitz, M., and J. H. Northrop. Isolation from beef pancreas of crystalline trypsinogen, trypsin, a trypsin inhibitor, and an inhibitor-trypsin compound. *J. Gen. Physiol., 19*, 991, 1936.

Landsteiner, K. *The Specificity of Serological Reactions.* Harvard University Press, revised edition, Cambridge, Mass., 1945.

Langmuir, I. Molecular layers. *Proc. Roy. Soc. London,* A., *170,* 940, 1, 1939.

Langmuir, I., and V. J. Schaefer. Activities of urease and pepsin monolayers. *J. Amer. Chem. Soc., 60,* 1351, 1938.

Lindegren, C. C., S. Spiegelman, and G. Lindegren. Mendelian inheritance of adaptive enzymes in yeast. *Proc. Nat. Acad. Sc., 30,* 346, 1944.

Linderstrom-Lang, K., and C. F. Jacobsen. The contraction accompanying enzymic breakdown of proteins. *C. R. Trav. Lab. Carlsberg, Sér. Chim., 24,* 1, 1941.

Lloyd, W., M. Theiler, and N. I. Ricci. Modification of the virulence of yellow fever virus by cultivation in tissues *in vitro. Tr. Roy. Soc. Trop. Med. Hyg., 29,* 481, 1936.

Loeb, J., and J. H. Northrop. On the influence of food and temperature upon the duration of life. *J. Biol. Chem., 32,* 103, 1917.

Loring, H. S. Solubility studies on purified tobacco mosaic virus. *J. Gen. Physiol., 23,* 719, 1940.

Luria, S. E., and M. Delbrück. Interference between inactivated bacterial virus and active virus of the same strain and of a different strain. *Arch. Biochem., 1,* 207, 1942.

Luria, S. E., M. Delbrück, and T. F. Anderson. Electron microscope studies of bacterial viruses. *J. Bact., 46,* 57, 1943.

Luria, S. E. Mutations of bacterial viruses affecting their host range. *Genetics, 30,* 84, 1945.

Madden, S. C., and G. H. Whipple. Plasma proteins. *Physiol. Rev., 20,* 194, 1940.

Marrack, J. R. *The Chemistry of Antigens and Antibodies.* 2nd edition. Sp. Rep. Ser. Med. Res. Coun., London, No. 230, 1938.

Meyerhof, O. Energy relationships in glycolysis and phosphorylation. *Ann. N.Y. Acad. Sc., 45,* 377, 1944.

Miller, G. L., and W. M. Stanley. Derivatives of tobacco mosaic virus. I. Acetyl and phenylureido virus. *J. Biol. Chem., 141,* 905, 1941.

Moore, D. H., S. C. Shen, and C. S. Alexander. The plasma of developing chick and pig embryos. *Proc. Soc. Exp. Biol. N.Y., 58,* 307, 1945.

Mudd, S. A hypothetical mechanism of antibody production. *J. Immunol., 23,* 423, 1932.

Muller, H. J. Pilgrim Lecture, London, November 1945 (in press).

Murphy, J. B. Transplantability of malignant tumors to the embryos of foreign species. *J. Amer. Med. Assn., 59,* 874, 1912.

Northrop, J. H. The inactivation of trypsin. III. Spontaneous inactivation. *J. Gen. Physiol., 4,* 261, 1922.

Northrop, J. H. The formation of enzymes. *Physiol. Rev., 17,* 144, 1937.

Northrop, J. H. Concentration and purification of bacteriophage. *J. Gen. Physiol., 21,* 335, 1938.

Northrop, J. H. *Crystalline Enzymes.* Columbia University Press, New York, 1939.

Northrop, J. H. Crystalline diphtheria antitoxin. *Proc. Amer. Phil. Soc., 85*, 13, 1941.

Northrop, J. H. Plastein from Pepsin and Trypsin. *J. Gen. Physiol., 30*, 377, 1947.

Orcutt, M. L., and P. E. Howe. Relation between the accumulation of globulins and the appearance of agglutinins in the blood of new born calves. *J. Exp. Med., 36*, 291, 1922.

Ostromuislenskii, I. I. Investigations of toxins and antitoxins. *Chem. Abstr., 10*, 214, 1916.

Pauling, L. Theory of the structure and process of formation of antibodies. *J. Amer. Chem. Soc., 62*, 2643, 1940.

Pauling, L., and D. H. Campbell. Manufacture of antibodies in vitro. *J. Exp. Med., 76*, 211, 1942.

Pedersen, K. O. *Ultracentrifugal Studies on Serum and Serum Fractions.* Almquist & Wiksells Boktryckeri AB, Uppsala, 1945.

Petrie, A. H. K. Protein synthesis in plants. *Biol. Rev. Cambridge Phil. Soc., 18*, 105, 1943.

Potter, V. R., and H. G. Albaum. Mechanism of hydrogen transport in animal tissues. VII. Inhibition by ribonuclease. *J. Gen. Physiol., 26*, 443, 1943.

Price, W. H. *Proc. Nat. Acad. Sci., 34*, 317, 1948.

Reiner, J. M., and Spiegelman, S. *Tech. Proc., 7*, 98, 1948.

Rivers, T. M. *Viruses and Virus Diseases.* Stanford University Press, Stanford University, California, 1943.

Robertson, T. B. The function of the lipoid in mitochondria. *Australian J. Exp. Biol. and Med. Sc., 3*, 97, 1926.

Roche, J., Y. Derrien and S. Mandel. Hétérogénéité des fractions eu- et pseudoglobuliniques des protéines du sérum de cheval et relargage par les sels neutres. *C. R. Soc. Biol. Paris, 138*, 634, 1944a.

Roche, J., Y. Derrien, and S. Mandel. Séparation de trois cristalbumines du sérum de cheval. *C. R. Soc. Biol. Paris, 138*, 665, 1944b.

Sabin, F. R. Cellular reactions to a dye-protein with a concept of the mechanism of antibody formation. *J. Exp. Med., 70*, 67, 1939.

Schoenheimer, R. *The Dynamic State of Body Constituents.* Harvard University Press, Cambridge, Mass., 1942.

Seastone, C. V., and R. M. Herriott. Immunological studies on pepsin and pepsinogen. *J. Gen. Physiol., 20*, 797, 1937.

Seifriz, W. A materialistic interpretation of life. *Phil. Sc., 6*, 266, 1939.

Sevag, M. G. *Immuno-Catalysis.* Charles C. Thomas, Springfield, Ill., and Baltimore, Md., 1945.

Sevag, M. G. Enzyme problems in relation to chemotherapy, "adaptation," mutations, resistance, and immunity. *Advances in Enzymology, 6*, 33, 1946.

Spiegelman, S., and M. D. Kamen. Genes and nucleoproteins in the synthesis of enzymes. *Science, 104*, 581, 1946.

Spizizen, J. Biochemical studies on the phenomenon of virus reproduction. *J. Infect. Dis., 72*, 212, 1943.

Stanley, W. M. Biochemistry and biophysics of viruses. In Doerr and Hallauer, *Handbuch der Virusforschung,* Verlag von Julius Springer, Wien, 1938a.

Stanley, W. M. The reproduction of virus proteins. *Amer. Naturalist, 72,* 110, 1938b.

Stanley, W. M. The biochemistry of viruses. *Ann. Rev. Biochem., 9,* 545, 1940.

Stanley, W. M. Chemical structure and the mutation of viruses. From *Virus Diseases.* Cornell University Press, Ithaca, N.Y., p. 35, 1943.

Svedberg, The. Protein molecules. *Chem. Rev., 20,* 81, 1937.

Sylverton, J. T., and G. P. Berry. Coexistent infections of individual cells by more than one filterable virus. *J. Bact., 32,* 356, 1936.

Taylor, A. E. On the synthesis of protein through the action of trypsin. *J. Biol. Chem., 3,* 87, 1907.

Taylor, A. E. On the synthesis of protamin through ferment action. *J. Biol. Chem., 5,* 381, 1909.

Taylor, H. S. (editor). *A Treatise on Physical Chemistry.* Two volumes. 2nd printing. D. Van Nostrand Co., New York, 1925.

Ten Broeck, C. The differentiation of trypsins by means of the anaphylactic test. *J. Biol. Chem., 106,* 729, 1934.

Traub, E., and C. Ten Broeck. Protective vaccination of horses with modified equine encephalomyelitis virus. *Science, 81,* 572, 1935.

Troland, L. T. Biological enigmas and the theory of enzyme action. *Amer. Naturalist, 51,* 321, 1917.

Wasteneys, H., and H. Borsook. The enzymatic synthesis of protein. *Physiol. Rev., 10,* 110, 1930.

Woodruff, A. M., and E. W. Goodpasture. The susceptibility of the chorio-allantoic membrane of chick embryos to infection with the fowl-pox virus. *Amer. J. Path., 7,* 209, 1931.

Yudkin, J. Enzyme variation in microörganisms. *Biog. Rev. Cambridge Phil. Soc., 13,* 93, 1938.

Yule, G. U. On the distribution of deaths with age when the causes of death act cumulatively and similar frequency distributions. *J. Roy. Statist. Soc. London, 73,* 26, 1910.

Zinsser, H. *Infection and Resistance.* 3rd edition. The Macmillan Co., New York, p. 141, 1923.

Zinsser, H. On the nature of virus agents. *Amer. J. Pub. Hlth., 27,* 1160, 1937.

Zinsser, H., J. F. Enders, and L. D. Fothergill. Immunity, principles and application in medicine and public health. 5th edition of *Resistance to Infectious Diseases.* The Macmillan Co., New York, 1939.

II. MOLECULAR MORPHOLOGY AND GROWTH

BY FRANCIS O. SCHMITT[1]

MOLECULAR morphology and growth are both very large subjects. Their combination encompasses a respectable portion of the content of biology. If we are to bring into sharp focus a few of the fundamental problems, to achieve a measure of synthesis and projection into profitable paths of future investigation, certain limitations of scope must be imposed.

One of the uniquely characteristic properties of protoplasmic systems is the structural organization by which the various energy-giving reactions are coupled and coordinated with the molecular effectors, the "machinery," of the cell. While the molecular morphology of the components which participate in cellular energetics, the enzymes, hormones, metabolites and so on is also highly critical, we shall concern ourselves chiefly with the macromolecular constituents of cell structure and in particular with the fibrous proteins.

In discussing a subject as broad as that of growth, it would be impossible to exclude so important a subject as the globular proteins, since many of the enzymes which preside over cellular energetics belong in this category. As we shall see, a consideration of fibrous proteins involves the globular proteins very closely. Unfortunately the space allotted here is too limited to permit a full discussion of the detailed structure of proteins generally.

I am not unmindful of the importance, for cell structure, of protein complexes which are not perceptibly fibrous in nature. Included with these are the various cytoplasmic particulates such as the microsomes, secretion granules, Nissl substance and so on. However, polarization and electron optical evidence indicates that the fibrous proteins form the basis of the molecular fabrics which constitute the structural matrix of the cell. Hence my own remarks will be limited to this aspect of molecular morphology.

With these apologies I shall proceed with a brief statement of the present state of knowledge about the fibrous proteins and offer some suggestions about fruitful fields for further study.

[1] Department of Biology, Massachusetts Institute of Technology, Cambridge, Massachusetts.

The Technical Background

The determination of protein structure requires a complete and exact knowledge of the amino acid composition and of the configuration of the amino acid residues. To what extent are the physical and chemical techniques so far developed adequate for this purpose and to what extent have these techniques been applied?

Until recently amino acid data were not only relatively unreliable but also incomplete with respect to the constituents present in small amounts. If proteins are to be dealt with stoichiometrically, the analytical data must be exact. Recent advances in amino acid analysis, particularly the stable isotope and microbiological and adsorption techniques, should help to fill many of the gaps in the cases where really pure preparations can be obtained. Unfortunately few fibrous proteins have been prepared in a state of purity approaching that which is required.

Fairly good technical and theoretical background exists for the determination of particle size and shape. This includes the methods of streaming double refraction, viscosity, diffusion and ultracentrifugation. But those methods yield appropriate information only when the preparation is truly monodisperse. In the case of myosin, for example, this criterion has not been fully met. The particles are fairly uniform in cross section but have highly variable lengths. Hence the measurements led to an average value for particle size, the larger sizes receiving greater weight, rather than to dimensions which could be ascribed to the myosin molecule.

Electron staining and shadow casting techniques have greatly enhanced the value of the electron microscope in the investigation of the structure of fibrous proteins. Resolution of the order of 25 A has already been achieved with favorable materials and it is probable that this may be further increased. Small-angle x-ray diffraction techniques permit the recording of reflections corresponding to periodicities of the order of hundreds of Angstrom units, thus widely overlapping the resolving power of the electron microscope. The electron microscope data assist greatly in the interpretation of the diffraction patterns of fibrous proteins such as collagen and paramyosin which manifest a large fiber-axis repeating period. In these cases fine structure within the main period can be demonstrated in the electron micrographs; this fine structure may be correlated with the x-ray data to deduce the structure.

Only a few of the fibrous proteins have been studied with all of the physical and chemical techniques mentioned. An attempt to show the state of our knowledge, or perhaps our ignorance, is presented in Figure 1. The individual rectangles represent eight methods of studying

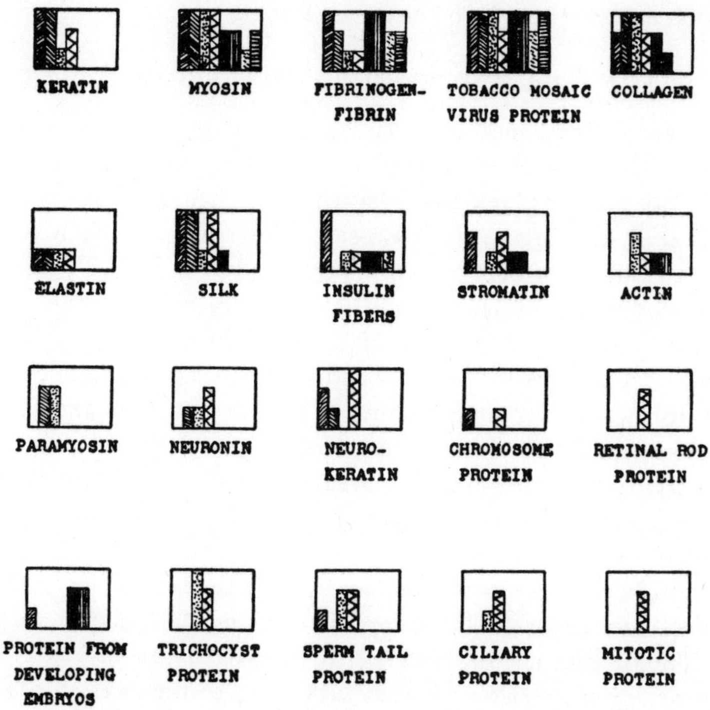

Figure 1. Rough appraisal of experimental data available on physical and chemical properties of fibrous proteins. The eight rectangles, each characteristically shaded, represent from left to right: amino acid analysis, x-ray diffraction, electron microscopy, polarized light microscopy, streaming double refraction, viscosity, diffusion and ultracentrifuge analysis. The height of each rectangle represents the relative amount of work published on the application of this method to the protein in question. Since this chart was prepared important new results have been published, particularly in the case of actin, myosin, and the actomyosin complex.

molecular structure. The height of each rectangle reflects the relative amount of work which, to the author's knowledge, has been reported on the subject. Of course, the mere fact that much work may have been done on a subject does not necessarily mean that crucial conclusions have been reached.

The figure is admittedly very rough, but the many gaps in our knowledge are obvious. There must be concentration both on breadth and depth. As more proteins are investigated the comparative aspect will help to crystallize our concepts. At the same time there should be heavy concentration on a few of the most favorable proteins which can be isolated in relative purity and which manifest a high degree of structural regularity. Lacking more direct and more sensitive methods, only thus may we approach the desired goal.

INTERNAL ARCHITECTURE OF FIBROUS PROTEINS

Astbury distinguishes two classes of fibrous proteins: the keratin-myosin-fibrin class, which gives the characteristic alpha and beta wide-angle x-ray patterns, and the collagen class. Astbury's general point of view is well summarized in his essay, "The Form of Biological Molecules," included in the set of essays presented to D'Arcy Thompson. The correlations which he has drawn have greatly helped to weave a common thread through the maze of assorted information which existed previously. The data do not yet warrant conclusions of detailed structure which are necessarily unique, but, as Astbury says, it must be "something like that." Until specimens are prepared which have sufficient regularity of structure to give a greater wealth of diffraction data, it is unlikely that the details of the folded alpha configuration will be deducible.

In certain of the fibrous proteins a large repeating fiber-axis spacing has been demonstrated by x-ray diffraction and electron microscopy. The value of this spacing and the intensities of the various diffraction orders are characteristic of each protein. Yet the true significance of this remarkable characteristic remains unknown. Astbury has attempted to interpret long spacings in terms of the amino acid composition along stoichiometric lines. The long spacing is regarded as the product of the number of amino acid residues in the molecule and the length, along the fiber axis, of each residue as indicated by the meridional diffractions of the wide-angle pattern. However, the data are not yet sufficiently complete to warrant this procedure except as a rough guide to thought. Moreover, Bear has demonstrated in the case of collagen that physical and chemical alteration may cause the long and short fiber-axis spacings to vary somewhat independently of each other.

While the wide-angle x-ray patterns reflect similarities of structure, the small-angle pattern is characteristic of each particular protein. Fairly well developed small-angle patterns have been obtained for α and β keratin, collagen and paramyosin. Myosin patterns have also been obtained (Bear), but as yet these lack detail. No x-ray long-spacings have been reported for fibrin, though Wolpers has observed a periodic

pattern in electron micrographs. The shafts of Paramecium trichocysts show very regular periodic structure in the electron microscope, but this object has not yet been studied with x-rays. In addition to the large fiber-axis periodicities, the fibrous proteins may also show large lateral separations, and these further assist the characterization of the protein. The information on these large periods is summarized in Table I.

TABLE I

Large Repeating Periods in Fibrous Proteins

Period	Repeating Period	
	Fiber-axis Period	Lateral Period
	A	A
β Keratin	95	34
α Keratin	198	82
Muscle (actomyosin?)	350–420	115
Paramyosin	720	200–300
Collagen	640	—
Trichocyst Protein	550 (2200)	—

Another approach to the problem is to make electron microscope and x-ray studies of fibrous proteins in which reactive groups have been altered chemically, as by deamination, deguanylation and so on. Such a program is currently being undertaken at M.I.T. on the protein collagen.

Finally, a word must be said about the nucleoproteins. In Astbury's study of mixtures of thymonucleic acid with the protamine, clupein, the x-ray patterns were very similar to those of nucleic acid alone. The only aspect which is stressed by Astbury is a fiber-axis period of 3.34 A which is characteristic of nucleic acid and is supposed to indicate the separation between nucleotide residues. Since this spacing is close to 3.5 A, the distance between amino acid residues in fully extended beta proteins, Astbury supposes a salt linkage between protein side chains and phosphoric acid residues of the nucleotides.

Fluxional methods have shown that nucleic acid occurs as large highly asymmetric particles of columnar shape. From polarization optics it is known that in these columnar particles the nucleotide residues are oriented with long axes perpendicular to the axis of the columns. The picture resembles stacks of coins. In each stack the coins are bonded to each other near the periphery of the stack. Unfortunately, no further detail can be furnished.

The structure and composition of virus nucleoproteins, particularly tobacco mosaic virus, has been closely studied. In this field the x-ray work of Bernal and Fankuchen and the chemical studies of Stanley and his associates are classical, but the published data do not yet permit a detailed description of the structure of the protein and nucleic acid constituents.

The Avery experiment, in which nucleic acid extracted from pneumococci has been likened to a free gene, has further heightened the interest in these substances. It is urgent, therefore, that the very meager information about the structure of nucleic acids and nucleoproteins be extended by further x-ray examination. Since, to the author's knowledge, no electron microscope studies have thus far been reported, the possibilities in this direction cannot be assessed.

The Continuous Polypeptide Chain Theory versus the Corpuscular Theory of Protein Structure

From the classical work of Emil Fischer it was natural to consider the polypeptide chain as the unit of protein structure. This view was adopted by those using x-ray methods, especially by Astbury. By stretching an alpha keratin fiber one was literally pulling the folded polypeptide chains into the straight, beta configuration. In addition it was supposed that these continuous polypeptide chains occur in well ordered semicrystalline regions and in poorly ordered amorphous regions; polypeptide chains coursing through both types of regions produce an anastomotic system.

Recent experiments of Bear and his associates, in which x-ray patterns are obtained from specimens which are tilted at various angles in the x-ray beam, support the view that collagen is essentially a one-dimensional structure. By this is meant that the linear elements have a thickness of only a few polypeptide chains and have indefinite extension longitudinally. Electron microscope evidence suggests that these units associate with each other laterally so that their axial periodicities are in phase, thus causing the entire fibril to appear cross-striated. It is even possible to dissociate these units by treating rat tail collagen with acetic acid, and to cause them to join up again, on elevation of the *p*H, so that cross-striated fibrils are again formed. In the case of collagen, therefore, the polypeptide chain theory seems most attractive at the present time.

Myosin, paramyosin and fibrin show wide-angle x-ray patterns similar to that of alpha keratin, from which it may be concluded that some similarity of structure of the polypeptide units exists. However, these proteins differ from keratin, at least superficially, in that they may be composed of particulate units. In such cases the continuous polypeptide chain description requires considerable qualification. Some examples of this class may be mentioned.

Fibrinogen molecules have been well characterized with respect to size and shape. They are long, thin spindle-shaped particles. The chemical change which occurs when these units combine to form fibrin fibers is very slight, making it improbable that the internal architecture suffers much alteration in the process. In this case the aggregation of the particles is not readily reversible.

Globular insulin molecules, which have relatively slight asymmetry, may be converted into a fibrous modification without denaturation in the strict sense of the term. The process is completely reversible, as Waugh has shown. This is a particularly strategic case because much is known about the composition and structure of the insulin molecule. This should greatly aid in investigating the alterations which occur when the globular molecules are recruited into fibrous arrays. Waugh has pointed out that the phenomenon requires the assumption that combination occurs at opposite poles of the molecule, otherwise only random gel-like aggregation, rather than filament formation, would result. It seems probable that similar considerations may apply to certain other particle-fiber transformations.

Finally the remarkable properties of the muscle protein, actin, may be mentioned. In distilled water these particles or molecules occur freely dissociated. The solution has relatively low streaming double refraction and viscosity. Merely increasing the ionic strength greatly increases the double refraction and viscosity due presumably to fibrous aggregation of the particles. Jakus and Hall have studied the phenomenon with the electron microscope and find that the filaments have widths of the order of 100 A and lengths up to several microns. Reduction of ionic strength causes the filaments to break up. The recent work of Szent-Györgyi, Cori, Price and others suggests that this type of reversible aggregation may be involved in muscle contraction.

The Mechanism of Fibrogenesis

The points just discussed bear importantly on the chemical and biological mechanisms of fibrogenesis.

In the case of fibers constructed of globular units it is easy to picture possible *intracellular* processes. Given a sufficient concentration of the appropriate globular protein molecules within the cell, their aggregation or disaggregation would depend on the local environment with respect to salts or activating enzymes. The actin-myosin cycle seems particularly attractive as a prototype. In this case the aggregation or disaggregation would depend on the presence of free adenosinetriphosphate (ATP) or similarly acting substance. The concentration of the latter depends, in turn, on the presence of ATP-ase and on the coupled reactions of intermediary metabolism. Thus would be exemplified the close relationship between the energy yielding reactions and structure formation in cells.

Little vision is required to suggest such possible mechanisms, but much experimentation is needed to isolate the proteins of cell structure and to determine their physical and chemical properties. Especially significant for the problem of growth is the fibrous system developed during the mitotic cycle. Existing theories of the formation of the spindle and astral fibers are well-nigh useless. It is necessary to tackle this problem with the techniques of modern protein and enzyme chemistry.

Little can be said about the mechanism of the formation of fibers of the collagen class, the components of which consist essentially of one-dimensional arrays. The fibers are presumably formed extracellularly. Do the cells produce a precursor, a procollagen, which is converted into collagen under the influence of the intercellular environment? Or do the fibroblasts produce the one-dimensional arrays of collagen which then pass into the extracellular space and there form fibers in a manner analogous to the reconstitution of fibers from an acetic acid solution of collagen by the addition of salt or elevation of *p*H?

To these questions there are as yet no satisfactory answers. Collagen is a good case to investigate, for the fibrils can be identified in the electron microscope.

Comparative Macromolecular Morphology

Reference has been made to a classification of the fibrous proteins according to two types, the collagen group and the group showing the characteristic alpha x-ray pattern, including keratin, myosin, fibrin, and paramyosin. Astbury has stressed the comparative aspect and has suggested that nature employs only a few general patterns of structure with major and minor variations on the main theme.

Aside from the value of comparative studies in suggesting clues about the structure of fibrous proteins, the matter has a biological significance as well. Thus a detailed comparison of the structure and composition of a particular protein, such as keratin, in the various phyla and classes may reveal relationships quite as fundamental as those of comparative anatomy and embryology.

Astbury and Rudall have begun such a study of the keratins. "Feather" keratin, which is of the beta variety, is found only in reptiles and birds; in other vertebrates the keratin is of the alpha type. In reptiles the mobile regions of the body are covered with the softer alpha keratin, while the remainder of the body is enclosed in the harder beta variety. Unfortunately the comparative studies on keratin have been made chiefly on the basis of wide-angle x-ray patterns.

Bear and his associates have been making a careful comparative x-ray study of the collagen type. All vertebrate collagens examined show the same characteristic wide-angle pattern as do also many invertebrate collagens, including those of representative mollusks, annelids, and echinoderms. Recently these investigators have found a similar pattern in gorgonin and spongin, thus furnishing evidence for the existence of this class of proteins in the coelenterates and porifera. The meager data that are available on the amino acid composition of these proteins in the lower forms indicate significant differences from that of vertebrate collagen. Bear's studies of the fiber-axis long spacings have added greatly to the detailed information necessary for comparative studies. It remains to be seen whether, with increasing phylogenetic complexity, this ubiquitous protein type also increases in regularity of molecular architecture or whether, even in the lowest forms, it exists already fully established. The shaft membrane of the trichocysts of Paramecium possesses a structure fully as regular as that of vertebrate collagen, to which it bears some resemblance in chemical properties, as shown by Jakus.

Will it turn out that the lowest organisms contain protein patterns as complex as those of the higher organisms; that such complexity of molecular structure is required for life as we know it? If so, what are the missing links between such structural proteins and the simpler organic compounds?

Macromolecular Ontogeny

The general problem of the chemical mechanism of the synthesis of proteins has been considered previously in this symposium. The problem may also be considered from the biological point of view. When in the embryological sequence are the various structural proteins elaborated? The embryologist can trace the steps in the differentiation of the nervous system, muscles, connective tissue and so on, from the earliest anlagen. But characteristic nerve fibers, myofibrils, and collagen fibers are differentiated considerably after the tissue is determined. When in these processes are the neuronin of nerve fibers, the myosin and actin of muscle fibers, and the collagen of connective tissue elaborated? Are these fibrous proteins produced full blown or are they formed from chemical "anlagen" or precursors? In other words, is there an ontogeny of the structural proteins or are they formed under the influence of templates preexisting in the embryonic cells, possibly even in the egg itself?

These are large questions which have been mostly academic in the past because the means did not exist by which they might be answered, but it is now possible to make a beginning in such investigations. For example, collagen fibrils can now be readily recognized in the electron microscope. It should be possible therefore to examine embryonic material at various stages of development to determine when and where collagen fibrils first make their appearance.

The technical aspects of the problem are by no means insurmountable and will no doubt yield if a determined effort is made. The possibility of laying the foundation for a new theoretical biology based on data obtained at the molecular level is worthy of the best collaborative efforts of biophysicists, biochemists and biologists. As the answers to these problems are found, a clearer knowledge of the phenomena of growth, normal and abnormal, must inevitably follow.

Molecular Ecology

The interaction of structural components with their environment is at the basis of general and developmental physiology. Such interactions depend not only on individual groups, such as particular amino acid side chains, but on the molecular reacting system as a whole. Molecular morphology should be studied not independent of or even consecutively with the dynamic aspects, but simultaneously where possible. This amounts to a study of molecular ecology, a term which I borrow from Dr. Paul Weiss.

In a sense this is implied in what has been said before. Thus certain fibrous proteins react spectacularly to changes in their environment.

Fibrinogen molecules are recruited into fibrin fibers in the presence of small amounts of thrombin. The actin-myosin system of muscle responds to changes of potassium ion concentration or to ATP in a dramatic manner. Such processes are readily appreciated in the light of the rich physiological background in blood clotting and muscle contraction.

But the general concepts of the lability of macromolecular complexes, their sensitivity to their chemical environment, is perhaps not fully appreciated by students of cellular physiology. The heritage of classical morphology conditions the investigator to static rather than to dynamic concepts.

Many examples might be cited to illustrate the point. Investigators of permeability phenomena in searching for the structure of the plasma membrane are prone to regard this organelle as a fixed mosaic of protein and lipid components. The small amount of evidence at hand indicates that the protein component consists of a feltwork of very thin protein filaments. Is it not possible that such proteins may be susceptible to changes in ionic environment, to enzymes or to materials like ATP, as is the case with the fibrous muscle proteins, and that such changes may result in alterations of the protein which may profoundly affect permeability properties? The protein of the erythrocyte envelope, stromatin, may be isolated and studied for such properties. Perhaps this would be a favorable system in which to test the suggested hypothesis.

The case of neurofibrils may also be mentioned. Clearly demonstrable by suitable cytological techniques, these structures have been regarded by physiologists as mere fixation artifacts. Polarization optics, however, has demonstrated the existence of longitudinal submicroscopic strands which form the framework of neurofibrils. This molecular lattice is extremely sensitive to its chemical environment. In what ways is this sensitivity involved in the propagation of the nerve impulse? In the normal maintenance, growth, and repair of the neuron? These questions cannot be answered until the neuronin complex is better characterized biochemically and structurally.

The chromosome is of course the final and best illustration of the importance of macromolecular ecology. On this structure is imposed the task not only of maintaining through countless duplications a rigid structural integrity—a process which physicists like Schrödinger find difficult to understand—but also of presiding over the multitudinous determinative processes of growth and development. The chromosome is capable of a determinative role not only by virtue of its specific intrinsic structure but also because of the very particular chemical environment provided by the orderly action of embryological organizers upon the reacting system as a whole. Further progress requires a frontal attack on the molecular structure of the chromosome with all the physical and chemical tools at our disposal. But the investigator who ventures in this difficult field must have a proper appreciation of the indeterminacies involved. These stem largely from the fact that chromatin removed from a nucleus for analytical study may be but a skeleton of the biological entity as it exists in the living cell. This should not deter him from making structural and chemical studies on chromosomal material isolated in whatever form is most convenient for the purpose, but the complexities imposed by the ecological factors should warn against oversimplified generalizations which might be derived from such partial systems.

III. PLANT GROWTH HORMONES

BY KENNETH V. THIMANN[1]

THE field of plant growth hormones, or auxins, has broadened out so enormously in recent years that I shall be able here to deal only with certain parts of it. The various applications to agriculture and horticulture, for instance, I will scarcely touch on, not because they are unimportant but because it seems preferable to center attention on the basic phenomenon of growth itself and how it is influenced by the hormones in plants.

One of the most striking features of the auxins is the multiplicity of their actions. Although they were discovered through their growth promoting effect on the Avena coleoptile, especially when applied to one side, and have been assayed in this manner ever since, it was soon made clear that they promote straight growth in many elongating stems and organs, i.e. enlargement along the longitudinal axis. Also, when such organs are slit and placed in a solution of an auxin they undergo differential growth due to greater elongation of the outer than the inner tissues. The auxins cause formation of roots on stems, they cause cell division in the cambium and the formation of callus at cut surfaces; on the other hand they inhibit the growth of buds, and they also inhibit both the elongation of roots and the formation of the abscission layer on leafstalks and fruitstalks. Further, they trigger in some way the swelling of the ovary into a fruit, as pollination does, so that by applying auxin in the absence of pollination a normal but seedless fruit is formed. Finally, in pineapple but not (so far as known) in other plants, they even cause flowering.

This multiple action is in strong contrast with that of some of the animal hormones which typically cause specific effects on specifically reactive tissues. The multiple action has led to the view, expressed especially by Went (1938) and more recently by Gautheret (1944), that auxin acts as a mobilizing agent, bringing specific hormones to the tissue at which it is applied. According to this view, root formation when auxin is applied to the base of a cutting would be due to the mobilization there of a specific root-forming hormone. Experiments of Went, of Bouillenne, and of Cooper have given indirect support to this view, but the support is not strong and there is evidence against it. For instance, the rooting of small cubes of potato tubers when given high

[1] Department of Biology, Harvard University.

auxin concentrations, or the rooting of fragments of *Helianthus* hypocotyls in tissue culture containing auxin, are difficult to explain on this basis. The influence of leaves in promoting the rooting of cuttings after auxin treatment, brought out especially clearly by Rappaport, was at first thought to be due to the contribution of the specific hormone by the leaves, but in the light of several more recent studies this effect is almost certainly nutritional. In at least one plant which is very hard to root with auxin treatment, the White Hibiscus studied by van Overbeek and coworkers (1946), the entire effect of the leaves could be replaced by nutrients, leaving no evidence for a hormonal factor.

Similarly the inhibition of lateral bud development by auxin applied at the tip was explained by the mobilization of bud growth factors at the point of auxin application, thus starving the buds of their essential factors. But inhibition has been shown to occur when auxin is applied to the lateral buds themselves; and although its mechanism is certainly not clear, it does seem to be essentially a direct effect of auxin on the lateral buds rather than a diversion of other hormones away from these buds. It is, however, possible in this case that the action is due to the intermediate formation of an inhibiting substance by the auxin. Then again, the inhibition of root growth, a phenomenon exhibited by all auxins and in very low concentrations, is observed on isolated roots and is definitely exerted on the root itself.

All these considerations very strongly support the idea which was tentatively put forward more than ten years ago (Thimann, 1935) that auxin controls some fundamental *master reaction* in the cell. From this reaction, growth in size differentation, or even inhibition may follow according to the abilities of the cells concerned and their supply of water and other factors. It remains true that during the occurrence of the actual growth process there is some mobilization of materials, i.e. of water and nutrients. But this is a part of the growth process itself, rather than its primary cause. Its part will be referred to later.

Before dealing with this fundamental master reaction and the evidence for it, it will be worthwhile to discuss one of the fields which has been most active in the last few years, namely the relation between *free and bound auxin.*

In many standard test objects growth is a simple function of the auxin applied, otherwise bio-assay would not be possible. From this it is but a short step to the concept that in the intact plant, or rather in a *part* of an intact plant, growth is a simple function of the auxin made available to that part. Of course nutrient materials have to be available too. This simple view had good success in explaining the facts so long as the auxin was determined by diffusion out into agar. The two-factor scheme given by Went and Thimann (1937) for many seedlings, and founded on diffusion experiments, worked very well. Perhaps the best case is furnished by Dolk's well-known study of geotropism in the coleoptile, where the excess growth of the lower side, which causes the geotropic curvature, was accounted for by the demonstration that 65% of the auxin (instead of 50% when vertical) went to the lower side. Many other organs have given essentially the same results.

The introduction of a new method, i.e. extraction of the auxin from the plant tissue with an organic solvent, has gradually undermined all this. The very first such experiments, on *Avena*, showed auxin all through the plant, including the parts which are no longer growing. This destroys the simple correlation between auxin and growth and

TABLE I

Auxin in Spinach Leaf Fractions

All samples extracted with ether 6 weeks at 2° after the treatment and subsequent acidification. Figures are units of auxin per mg. dry weight.

FRACTION	TREATMENT				
	None	Incubated 24 hours at 37°: at *p*H 8	at *p*H 10.5	at *p*H 10 with chymo-trypsin	Boiled at *p*H 10
Chloroplasts	0.21	3.54	0.07	*9.4*	0.16
Globulin I plus II	0.54	2.00	0.22	*21.0*	0.26
Albumin	—	2.43	0.80	1.1	0.50

introduces a distinction between free auxin, which diffuses readily into agar, and bound auxin, which does not. As Went has shown (1942), it is only the former which is redistributed under the influence of gravity. Extension of the extraction method to other material has raised many problems. Let us consider first the case of green leafy tissue. In *Lemna*, for instance, auxin is set free into ether continuously but very slowly. From dried material no liberation occurs, but on wetting it re-starts. The auxin is apparently being set free by hydrolysis (Thimann, Skoog, and Byer, 1942). Proteolytic enzymes, especially chymotrypsin, liberate the auxin rapidly, so that we have here an auxin-protein. Using chymo-trypsin we have attempted to locate this auxin-protein in the leaves of spinach, but there is more than one location. Some two-thirds of the auxin is associated with a globulin in the cytoplasm, but there is also much auxin in thoroughly washed intact chloroplasts. The other protein preparations do not contain large amounts. The table summarizes one such set of extractions. Bonner and Wildman (1946), using hydrolysis by alkali, found all the auxin in a globulin, a very interesting compound comprising three-quarters of the cytoplasm protein and having properties, including phosphatase activity, similar to those of myosin. Alkaline hydrolysis does not liberate the chloroplast auxin, which explains our inability to extract auxin from *Lemna* by alkali. Avery and coworkers (1945) have reported the extraction of auxin from leaves of cabbage by boiling in alkali; but this cannot be a general phenomenon, for Miss Mes, with tomato leaves (unpublished work), finds as we do with spinach and *Lemna* that there is no liberation of auxin by alkali.

Very little auxin is liberated by simple short extraction of leaves with ether, i.e. very little is free in the leaf. If this relatively large store of auxin in protein form were available to the plant for growth, it would be expected that leaves would continue growing much longer. One might expect giant cells and other evidences of high auxin concentration; but, further, one would expect to find the auxin very readily set free under natural conditions. Nevertheless many extensive experiments which we performed with different kinds of autolysis or action of autogenous enzymes succeeded only in showing a very small liberation; no large fraction of the auxin was ever obtained even after many days' autolysis.

A measure of the free auxin in the tissue is given by the results of short-time extraction with ether, or in some cases, extraction of boiled tissue with ether. Van Overbeek *et al.* (1945) have given evidence that the free auxin is essentially extracted in two hours. From our own data, based on twenty-four hours' extraction, the following figures may be given for the percentage of the total auxin given up in this way, i.e. the presumptive "free" auxin:

Fronds of *Lemna*	2%
The alga *Ulva*	2%
Avena roots	25%
Bean nodules	87%
Avena coleoptiles	Almost 100%

The contrast between the two extremes is very striking, for the auxin in nodules and coleoptiles is very largely in the free state while that in the green tissue is overwhelmingly bound. It is difficult to resist the conclusion that this protein-bound auxin of leaves is *not* a reserve and is (mainly at least) unavailable to the plant for growth.

Seeds, however, show a very different picture. When dry seeds of *Avena* or of corn are dampened, auxin is very rapidly set free. Heating with alkali liberates auxin very rapidly in this material, as shown by Avery and coworkers (1941). It is now clear that this auxin, too, is protein-bound, and Gordon (1946) has prepared pure proteins from wheat which liberate auxin on hydrolysis with alkali or acid. We know that seeds deliver to the seedling a precursor which the seedling tip converts to auxin. While this mobile precursor, partially purified by Berger and Avery (1944), does not itself seem to be a protein, it is clear that the auxin-protein of seeds is a true reserve or auxin-source, available to the plant for growth.

A word of warning is in place here. Tryptophane on heating with alkali forms a small amount of indoleacetic acid; casein and other proteins do the same. This phenomenon probably cannot account quantitatively for the auxin liberated by alkali from these seed proteins, but it certainly enters in. The fact that some seedlings are able to convert both tryptophane and tryptamine to free auxin should be remembered also. The demonstration by Haagen Smit, and later by Avery et al., and by Larsen and van Overbeek, that indoleacetic acid is no "heteroauxin" but a true plant auxin is of importance here.

Besides the above instances, there is still another type of behavior. When grass-stalks are laid horizontal, the nodes, which have long ceased growing, begin to grow on the lower side and this produces geotropic curvature of the stalks. Many years ago Schmitz showed by diffusion that this is due to the new production of auxin in these nodes; here gravity, instead of merely redistributing the auxin, as in the coleoptile, causes renewed *production* of auxin. Now van Overbeek *et al.* (1945) have shown in sugar cane that this new production is at the expense of bound auxin in the node; gravity in some way initiates the conversion of the bound to the free form. This phenomenon of gravitational release of auxin may be connected with another curious phenomenon. In pineapple (but not in other plants) application of auxin to the growing tip causes formation of flowers. Van Overbeek and his coworkers (unpublished) have now found that placing the pineapple plant horizontal has the same effect. This remarkable result may be due to liberation of bound auxin, but could of course be merely the result of geotropic accumulation of enough extra auxin on one side to start the flowering process.

In summary we may conclude that growth depends not only on the free auxin present but also on the ability of the tissue to convert bound auxin into the free form. The bound auxin is itself of many types: (1) that of the seed and perhaps the ovary which is made available for growth, (2) that of the leaf which is not, and (3) that of the main axis, which may occupy an intermediate position, while (4) that in the coleoptile appears much more loosely bound than the types discussed above.

I come now to the *mechanism of the action* of auxin. One of the greatest values of our knowledge of the auxins is that they provide a tool for the analysis of the growth process itself. I will try to outline briefly the progress made in this direction, but first I want to make the point that all this evidence that auxin can combine with proteins, and in many different ways, has another significance. From studies of the structure of the many substances active as auxins, Koepfli, Went and I have earlier deduced (1938) that there is a certain spatial relationship between the various parts of the molecule which is essential for activity. This suggests that the auxin has to combine with another molecule whose spatial structure is fixed, such as a protein, for example. This leads naturally to the idea that the auxin is a coenzyme for some enzyme system which controls growth.

It has been known for a long time that growth of the coleoptile will not take place anaerobically, and Bonner showed that growth is inhibited by cyanide to the same degree as respiration is. This suggests that growth is in some way linked to respiration, but it does not provide a mechanism for studying it. However, Commoner and I (1941) subsequently found that various known inhibitors of dehydrogenase systems also inhibit growth. This led to a study of the whole relationship between enzyme inhibitors and growth.

The technique was the straight growth of isolated coleoptile sections, which grow little in water or sugar alone, more in auxin, and considerably more in the combination of auxin and sugar. Growth was in shallow layers of solution by placing the sections on the teeth of combs.

Of all the inhibitors studied, iodoacetate was one of the most effective, giving complete inhibition at about 3.10^{-5} molal, an exceedingly low concentration for most enzyme inhibitions. Under these conditions the respiration of the sections was very little reduced. We have here not an effect on the general vitality or integrity of the cells, but specifically on the growth process itself. Furthermore this inhibition of growth is removed by malate and other four-carbon acids, and also by pyruvate and (as we have found later) isocitrate. Also, the growth of the uninhibited sections in auxin and sugar is clearly promoted by four-carbon acids after some hours. It is clear then that growth is not a function of respiration as a whole but of some specific part of it, i.e. of a respiratory system or at least a hydrogen-transferring system, involving those organic acids which participate in the Krebs cycle.

Confirmation of this came from the study of protoplasmic streaming in the coleoptile. Mrs. Sweeney and I (1938, 1942) studied this in the epidermal cells, at first by direct observation with a stop watch and later with a special device which provides an artificial comparison stream whose rate can be adjusted to synchronize with that of the protoplasm particles. The rate of streaming is accelerated by auxin. The reaction is closely related to growth because:

(a) it takes place before visible growth occurs, being clear within a few minutes,
(b) oxygen is required, as in growth,
(c) sugar is needed for the effect to last, just as in growth,
(d) the dependence on auxin concentration is parallel to that of growth,
(e) the increased rate is prevented by iodoacetate,
(f) the auxin effect is promoted by malate, especially in old or starved coleoptiles, whose sensitivity to auxin has been greatly reduced.

The parallel is thus very complete; the effect of auxin on streaming is almost undoubtedly a part of its effect on growth.

If acceleration of growth and acceleration of streaming are due to a dehydrogenation process, the effect of auxin on respiration should be interesting.

All investigators are agreed that auxin does not raise the oxygen consumption of coleoptile tissue in buffer solution; but we found that in sugar (sucrose and fructose) after a few hours' soaking, auxin does produce a rise of some 10% in oxygen consumption. Furthermore, if the tissue is soaked in malate or fumarate, the rise is larger (about 22%). This effect of auxin on respiration in the presence of malate closely parallels the effect on growth.

The effect of auxin on respiration in presence of sugar has recently been confirmed by Avery and Berger, though not however the influence of malate, which evidently needs further study to elucidate the conditions under which it occurs.

The inhibition of growth by iodoacetate and its promotion by four-carbon acids has been confirmed also for the growth of intact seedlings by Albaum and coworkers (1941, 1943). They found certain differences in the time relations, however, which may be due to the use of whole seedlings with accompanying complications of transport, inactivation in the plant, and varying uptake by the roots, or may be due as they suggest to changes in the enzyme systems with age of the seedlings.

Recently Bonner and Wildman have reported (1946) that fluoride inhibits growth, and like iodoacetate the inhibition can be exerted without any marked reduction in total respiration. They were led to this by their finding (mentioned above) that the auxin-protein in the cytoplasm of leaves contains a phosphatase and no other enzyme—a very suggestive fact. Fluoride and iodoacetate both act on the phosphorylating breakdown of glucose, in fact on directly subsequent reactions, though it is true that in the extracted systems so far studied they act only in higher concentrations than are effective on plant growth. We know from the work of James and his coworkers at Oxford that sugar breakdown in leaves follows closely (though not identically) the chain of reactions worked out in yeast and elsewhere. It is tempting therefore to postulate that both interfere with growth through the sugar-phosphorylating chain, and this would then be the system that delivers energy for growth. The four-carbon acids might participate by allowing a phosphorylation to occur at the expense of oxidizing malate, etc., as described for succinic, pyruvic, and ketoglutaric acids by Colowick *et al.* (1940, 1941) and by Ochoa (1943, 1944).

In attempting to get some evidence on this we have carefully compared the action of these two poisons on growth. Some time ago I noticed that iodoacetate, although a growth inhibitor, increases the auxin curvatures in the pea test, although it inhibits straight growth of the same pea seedling internodes at the same cencentration. The curvature may be increased by 50% while growth is decreased by some 50%. This characteristic is shown only very slightly, if at all, by fluoride. The highest nontoxic concentrations of both give opposite effects. Also, W. Bonner and I have found that the inhibition by fluoride is not released by the organic acids, though it can be released by certain hexose phosphates. Our tentative conclusion is that iodoacetate and phosphate inhibit growth in different ways or through different systems. The one affected by iodoacetate is presumably of sulfhydryl nature, and there is other evidence to support this.

At this point it is useful to digress for a moment to consider other organisms. If auxins influence growth through such fundamental systems as the Krebs cycle and the phosphorylation of hexose, we should not expect their action to be limited to higher plants. It is important in this connection that the long disputed question whether they control the growth of algae has recently been answered. Algeus (1946) has shown that indoleacetic acid does in fact promote cell division in green unicellular forms and has given an explanation of why up to now the results have been conflicting. One major factor is the strong influence of *p*H on entry of auxin acids into the cell, and the other is that the light necessary for continued growth of the algal cultures produces decomposition products of the auxin which are toxic. If these points are recognized and controlled, clear-cut growth promotion is found. Also Dr. Dubos has informed me that there are in some bacteria very marked effects of auxin on growth. This, together with some new evidence (unpublished) about an influence of auxin on muscle tissue in animals, justifies the view that the action is exerted via systems which are of quite general occurrence.

Lastly we must ask how this purely respiratory process can bring about growth. In the last analysis growth in plants means the intake of water. One way in which this may be brought about is through an effect on the accumulation of solutes. We know from the work of the California school that solute accumulation requires oxygen and consumes sugar and that active protoplasmic streaming is characteristic of cells which can accumulate. Following the demonstration by Reinders (1942) that auxin promotes water uptake by potato slices, Commoner and coworkers (1943) have shown that auxin even causes the uptake of water from a solution which is hyper-tonic, i.e. a solution which without auxin would cause wilting. This intake of water closely parallels other effects of auxin on growth, and it is promoted by potassium ions in much the same way as we earlier found the growth of coleoptiles in ordinary auxin solutions to be increased by low concentrations of KCl.

Very tentatively, therefore, one may propose the following steps for the simplest kind of growth of isolated plant parts immersed in solutions:

1. Auxin enters the cells with great rapidity, accelerating the rate of protoplasmic streaming.

2. This acceleration causes increased intake of solutes.
3. The energy for the accumulation of solutes and for the streaming is furnished from a sugar breakdown which involves phosphorylation and in which the Krebs cycle participates; one of the key enzymes here is of the sulfhydryl type.
4. The intake of solutes is at once accompanied by intake of water; growing cells do not have any higher osmotic pressure than non-growing ones.
5. The acceleration of streaming causes also a difference in the rate of wall deposition, i.e. a change in plasticity.
6. The increased water intake coupled with the change in wall properties means increased cell size—in other words, *growth.*

In this way one can begin to see how auxin can have the multiple effects mentioned at the outset. For a cell may not be in a condition to enlarge directly, due either to secondary wall formation or to its unfavorable situation in other tissues, yet the increased streaming rate and increased uptake of solutes may in some way activate numerous processes, causing cell division and leading to root initials and the like. Such a tissue would accumulate organic solutes too, thus providing some basis for the mobilization phenomena which, as was stated above, are probably secondary and not causal.

This scheme is still largely hypothetical but there is enough circumstantial evidence to make one believe that it is at least somewhere near the truth.

Note Added in Proof

The critical role of a sulfhydryl enzyme in growth has been confirmed (Thimann, K. V., and W. D. Bonner, Jr., *Amer. J. Bot., 35,* 271-281, 1948; and in press). The conversion of tryptophane to auxin has been clarified (Wildman, S. G., M. Ferri, and J. Bonner, *Arch. Biochem., 15,* 131-144, 1947) and shown to occur in the living leaf. In spite of these and many other recent results, the exact biochemical role of auxin in growth remains elusive.

REFERENCES

Albaum, H. G., and B. Commoner. *Biol. Bull., 80,* 314-323, 1941.
Albaum, H. G., and B. Eichel. *Amer. J. Bot., 30,* 18-23, 1943.
Algeus, S. *Bot. Notiser* (Lund), 129-278, 1946.
Avery, G. S., J. Berger, and B. Shalucha. *Amer. J. Bot., 28,* 596-607, 1941.
Avery, G. S., J. Berger, and R. O. White. *Amer. J. Bot., 32,* 188-191, 1945.
Berger, J., and G. S. Avery, Jr. *Amer. J. Bot., 31,* 11-19, 199-203, 1944.
Bonner, J., and S. Wildman. *Growth,* Supplement, Vol. *10,* 51-68, 1946.
Colowick, S. P., M. S. Welch and C. Cori. *J. Biol. Chem., 133,* 359-373;

641-642, 1940; Colowick, S. P., H. M. Kalckar, and C. Cori. *Ibid.*, *137*, 343-356, 1941.

Commoner, B., and K. V. Thimann. *J. Gen. Physiol.*, *24*, 279-296, 1941.

Commoner, B., S. Fogel, and W. H. Muller. *Amer. J. Bot.*, *30*, 23-28, 1943.

Gautheret, R. J. *Rev. Cyt. Cytophysiol. Vég.*, *7*, 45-185, 1944.

Gordon, S. A. *Amer. J. Bot.*, *43*, 160-169, 1946.

Koepfli, J. B., K. V. Thimann, and F. W. Went, *J. Biol. Chem.*, *122*, 763-780, 1938.

Ochoa, S. *J. Biol. Chem.*, *151*, 493-501, 1943; *155*, 87-97, 1944.

Overbeek, J. van. *Ann. Rev. Biochem.*, *13*, 631-666, 1944.

Overbeek, J. van, C. D. Olivo, and E. M. S. de Vazquez. *Bot. Gaz.*, *106*, 440-451, 1945.

Overbeek, J. van, S. A. Gordon, and L. E. Gregory. *Amer. J. Bot.*, *33*, 100-107, 1946.

Reinders, D. E. *Rec. trav. bot. Néerl.*, *39*, 1-140, 1942.

Sweeney, B. M., and K. V. Thimann. *J. Gen. Physiol.*, *21*, 439-461, 1938; *25*, 841-854, 1942.

Thimann, K. V., F. Skoog, and A. C. Byer. *Amer. J. Bot.*, *29*, 598-606, 1942.

Thimann, K. V. *Proc. Kon. Akad. Wetenschap.* (Amsterdam), *38*, 896-912, 1935.

Went, F. W. *Plant Physiol.*, *13*, 55-80, 1938.

Went, F. W. *Plant Physiol.*, *17*, 236-249, 1942.

Went, F. W., and K. V. Thimann. *Phytohormones.* New York, Macmillan, 1937.

IV. UNIDENTIFIED VITAMINS AND GROWTH FACTORS

BY KARL FOLKERS[1]

THERE are many investigations in progress which are concerned with the elucidation of new vitamins and growth factors. These investigations include nutritional experiments with animals, poultry, microorganisms, insects and other forms of life. Because of the magnitude of the literature on unidentified vitamins and growth factors, it seemed best for this review to select the results of certain investigations for discussion.

Experience has shown that the initial phases of certain nutritional investigations can be quite independent and diverse, and yet the final phases may be the same.

Turning to the past, an excellent example of initially diverse nutritional investigations was the independent researches on coenzyme R which was described as a growth and respiration promoting factor for *Rhizobium trifolii*, on biotin which is a yeast-growth factor, and on vitamin H which was a factor for rat nutrition (Table I).

In the final phase of these investigations, it was concluded that these three independent researches on the nutrition of *Rhizobium trifolii*, yeast, and rats were concerned with a single entity—biotin.

Turning to the present, it is desirable to discuss first folic acid and possible chemically related factors. These factors appear to deserve mention first for several reasons. The chemistry of the first members of the group to be elucidated is now becoming available. All the members of this group are receiving much study at present in many laboratories. These factors as a group typify again the development of independent, but frequently converging lines of nutritional studies.

These factors were introduced into the literature by such designations as vitamin M which is a factor for cytopenia in monkeys; norite eluate factor for the growth of *L. casei*; vitamin Bc for a microcytic hyperchromic type of anemia in chicks; factor R for chicks; folic acid for the growth of *S. lactis* and *L. casei*; liver *L. casei* factor; *Streptococcus lactis R* factor; vitamins B_{10} and B_{11} for chicks; fermentation *L. casei* factor; and the vitamin Bc conjugate. The early papers on these factors (Table II) appeared during 1938-1945.

[1] Research Laboratories, Merck & Co., Inc., Rahway, N.J.

TABLE I

Coenzyme R, Biotin and Vitamin H

Coenzyme R
(growth and respiration promoting effects of extracts for *Rhizobium trifolii*; Allison, Hoover and Burk [1933])

Biotin
(yeast-growth factor; Kögl [1935])

Vitamin H
(rat nutrition György [1939])

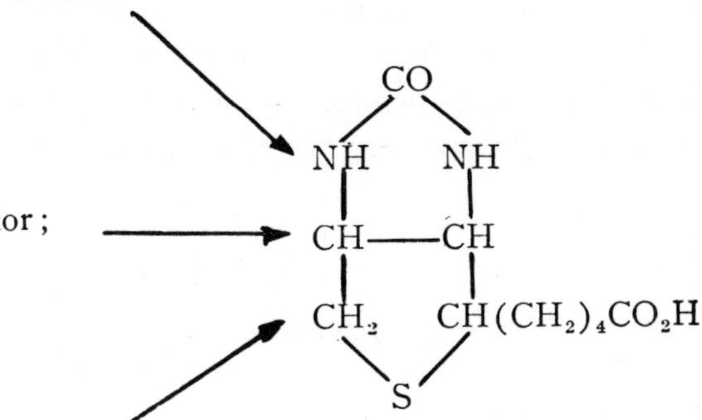

Biotin

TABLE II

Folic Acid and Possible Chemically Related Factors

Vitamin M	for cytopenia in monkeys	Day, Langstron, and Darby (1938)
Norite eluate factor	for growth of *L. casei*	Snell and Peterson (1939)
Vitamin Bc	for anemia (macrocytic hyperchromic type) in chicks	Hogan and Parott (1939)
Factor R	for chicks	Schumacher, Heuser, and Norris (1940)
Folic acid	for growth of *S. lactis* and *L. casei*	Mitchell, Snell, and Williams (1941)
Liver *L. casei* factor		Stokstad (1943)
Streptococcus lactis R factor		Keresztesy, Rickes, and Stokes (1943)
Vitamins B_{10} and B_{11}	for chicks	Briggs, Luckey, Elvehjem, and Hart (1943)
Fermentation *L. casei* factor		Hutchings, Stokstad, Bohonos, and Slobodkin (1944)
Vitamin Bc conjugate	for chicks	Pfiffner, Calkins, O'Dell, Bloom, Brown, Campbell, and Bird (1945)

During these independent researches on the nutrition of monkeys, chicks, and microorganisms, there were indications that some of the studies were concerned with a single entity, but were of unproven relationship during the early work.

The last few months have marked the beginning of the clarification of the chemical structural relationships between some of the factors which appear to be chemically related to folic acid.

Recent papers presented at the New York Academy of Sciences and published in *Science* showed that the liver "*L. casei* factor" is *N* [4{([2-amino-4-hydroxy-6-pteridyl]methyl) amino}-benzoyl] glutamic acid. The name pteroylglutamic acid has been suggested as a more convenient designation for this new vitamin. These elucidations of

TABLE III

Pteroylglutamic Acid

Liver "*L. casei*" factor
Stokstad (1943)

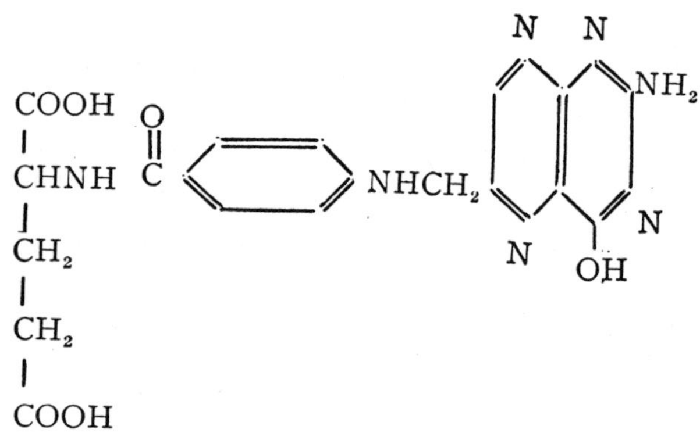

Pteroylglutamic acid

Vitamin Bc
Hogan and Parrott (1940)
Isolation, Pfiffner and coworkers (1943)

structure and synthesis studies were carried out in the Lederle Laboratories and the Calco Chemical Division of American Cyanamid.

A comparison of the synthetic pteroylglutamic acid with natural vitamin Bc in the Parke-Davis Laboratories showed that these two factors were identical (Table III).

Structural studies showed that the fermentation *L. casei* factor consists of three glutamic acid moieties in peptide linkage and attached through the free amino group of the tripeptide to the carboxy group of pteroic acid. The name pteroic acid was suggested for the acid:

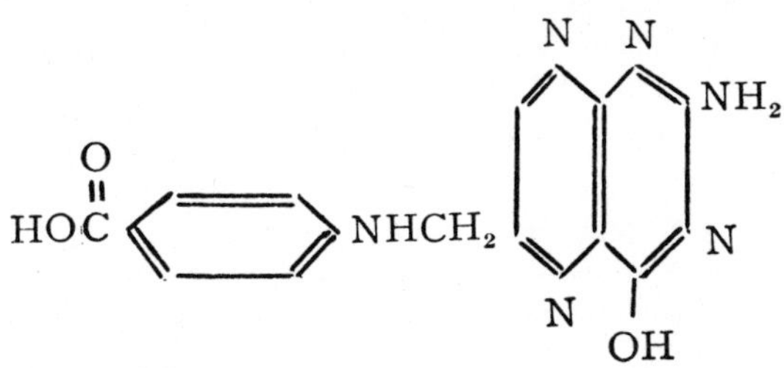

The vitamin Bc conjugate consists of seven glutamic acid moieties as a heptapeptide attached similarly to pteroic acid. The nature of the peptide linkages in the glutamic acid peptides has not been described as yet (Table IV).

The isolation of a polypeptide containing p-aminobenzoic acid from yeast has been described. Analytical data showed that this peptide had a free aromatic amino group, ten-eleven L-glutamic acid moieties and an unknown amino acid moiety in a unit of the molecule.

A functional relationship of the folic acid conjugates to this peptide of p-aminobenzoic acid is not established but is an obvious possibility.

The possible existence of other conjugates of pteroic acid containing two, four, five, etc., moieties of glutamic acid appears even more possible after the announcement of this yeast peptide (Table V).

It is evident that several factors which appear to be related chemically to pteroylglutamic acid remain for further chemical characterization. These factors are the norite eluate factor, folic acid, *Streptococcus lactis R* factor, vitamins B_{10} and B_{11}, factor R, and vitamin M.

It does appear possible that some of these materials might have contained one or more of the established pteroic acid conjugates of glutamic acid or conjugates with a number of glutamic acid moieties other than 3 or 7 (Table VI).

TABLE IV

Fermentation "*L. Casei*" Factor and Vitamin Bc Conjugate

Fermentation "*L. casei*" Factor
Isolation (1944)
Hutchings and coworkers →

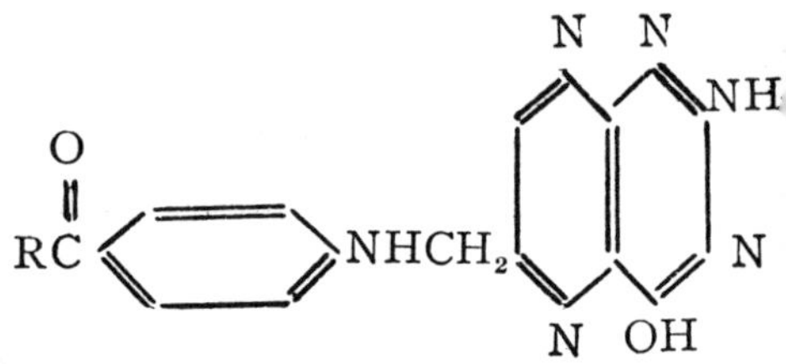

R = three glutamic acid moieties

Vitamin Bc Conjugate
Isolation (1945)
Pfiffner and coworkers →

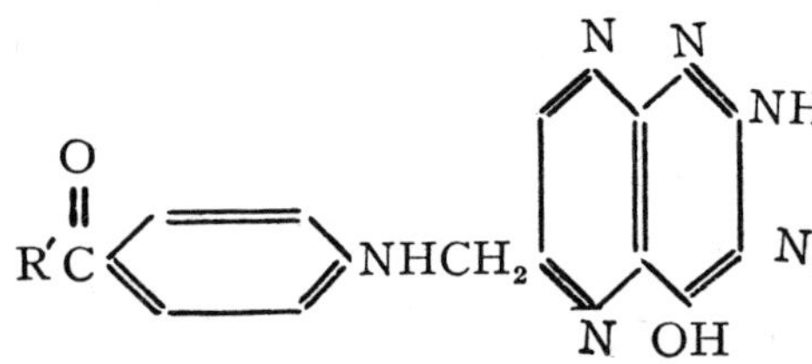

R′ = seven glutamic acid moieties

There are still other factors which may be related to pteroylglutamic acid.

In their attempts to prepare a concentrate of folic acid from by-products of commercial liver extract, Barton-Wright, Emery, and Robinson found that there were chloroform-soluble fractions which were active for the growth of *L. helveticus* and *S. lactis*.

Daft and Sebrell have studied the anemia and granulocytopenia which develop in rats on a pantothenic acid-deficient diet but with added *L. casei* factor. Treatment with whole dried liver was more effective than treatment with pantothenic acid. A factor or factors in the liver were indicated.

The significance of these observations and a possible relationship to pteroylglutamic acid is not yet clear (Table VII).

Because of abundant evidence on the antianemic properties of pteroylglutamic acid, or vitamin Bc, on experimental animals, it is not surprising that pteroylglutamic acid showed antianemic properties in man.

TABLE V

Isolation of a Peptide of p-Aminobenzoic Acid from Yeast
(Ratner, Blanchard and Green [1946])

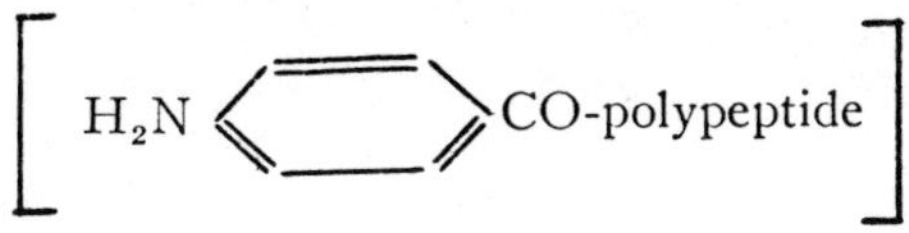

10-11 *l*-glutamic acid residues
1 unknown amino acid

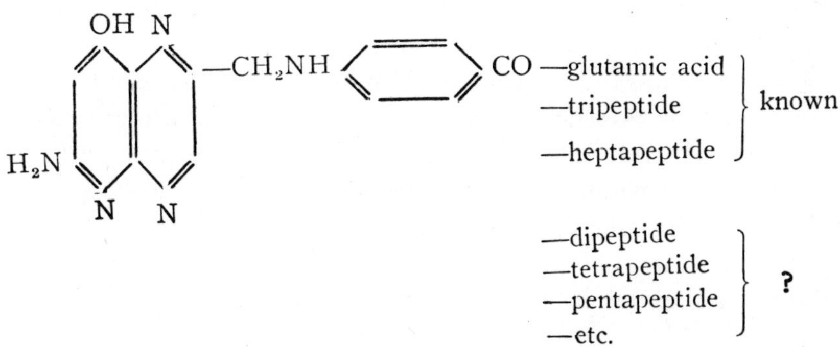

TABLE VI

Factors Possibly Related Chemically to Pteroylglutamic Acid

Norite eluate factor	Vitamins B_{10} and B_{11}
Folic acid	Factor R
Streptococcus lactis R factor	Vitamin M

Spies and his associates reported on nine cases of macrocytic anemia in relapse which showed a positive reticulocyte response and an increase in the red blood cell count after treatment with pteroylglutamic acid.

Darby and Jones reported satisfactory remissions when two patients with sprue were similarly treated, and Spies reported satisfactory remissions of tropical sprue.

Satisfactory remissions for several cases of Addisonian pernicious

TABLE VII

Other Factors Related (?) to Pteroylglutamic Acid

Growth factors for *L. helveticus* *S. lactis*	Chloroform soluble fractions from liver folic acid concentration process	Barton-Wright, Emery and Robinson, (1945)
Unidentified factor(s) in liver effective in blood dyscrasias in rats	Rats on pantothenic acid deficient diet plus *L. casei* factor	Daft and Sebrell (1945)

anemia and pernicious anemia of pregnancy have been reported by Moore and associates and by Amill and Wright.

Other such clinical observations have now been reported. However, as Moore and his associates pointed out, there is no clinical information as yet as to the effect of pteroylglutamic acid on the neurological manifestations of pernicious anemia (Table VIII).

The use of liver therapy for the treatment of pernicious anemia was introduced by Minot and Murphy in 1926.

Since pteroylglutamic acid has been isolated from liver, it is desirable

TABLE VIII

Pteroylglutamic Acid as an Antipernicious Anemia Substance

Macrocytic anemia in relapse (9 patients)	Positive reticulocyte response increased red blood cell count	Spies and associates (1945)
Sprue (2 patients)	satisfactory remissions	Darby and Jones (1946)
Tropical Sprue	satisfactory remissions	Spies (1945)
Addisonian pernicious anemia (2 patients)	satisfactory remissions	Moore and associates (1945)
Pernicious anemia of pregnancy (1 patient)	satisfactory remissions	Moore and associates (1945)
Addisonian pernicious anemia (6 patients)	satisfactory remissions	Amill and Wright (1946)

to know whether the effectiveness of commercial liver preparations is due to its content of pteroylglutamic acid. Numerous microbiological assays have been made with *L. casei* for the content of pteroylglutamic acid in liver preparations. For example, Clark reported that commercial parenteral liver extracts contained only 0.02-3.67μ of pteroylglutamic acid per unit of liver extract.

It was found that one U.S.P. unit of a certain liver extract when administered daily for ten days was equivalent to only 3.8μ of pteroylglutamic acid. A dose of 20 mg. of pteroylglutamic acid daily for ten days is 200,000μ. Alexander reported that only 0.035 mg. of a liver fraction was required to show antipernicious anemia activity.

Furthermore, Stokstad and Jukes have recently described the failure of vitamin Bc conjugase to liberate appreciable pteroylglutamic acid from a parenteral liver extract.

Thus it appears that the antipernicious anemia factor in liver is not pteroylglutamic acid, although the latter has antipernicious anemia activity (Table IX).

TABLE IX

Treatment of Pernicious Anemia by Liver Therapy

Liver therapy, Minot and Murphy (1926)
Commercial parenteral liver extracts contain 0.02-3.67 μ/unit, Clark (1945)
1 U.S.P. unit liver extract daily for ten days ≡ 3.8 μ pteroylglutamic acid
20 mg. pteroylglutamic acid daily for ten days ≡ 200,000 μ, Moore and associates (1945)
0.035 mg. Liver fraction active, Alexander

Parenteral liver extract $\xrightarrow[\text{dried chicken pancreas}]{\text{“vitamin Bc conjugase”}}$ no appreciable liberation of pteroylglutamic acid, Stokstad and Jukes (1946)

Castle and his associates in 1944 studied the effect of the daily administration of various substances together with neutralized normal human gastric juice on several cases of pernicious anemia. The data on two typical cases are summarized in Table X. In addition to an adequate amount of "vitamin-free" casein and all the known B vitamins, xanthopterin and a folic acid concentrate were also administered. It was concluded that folic acid had no effect upon the red blood cell count or the reticulocyte peak. However, when beef muscle was used as a sup-

TABLE X

Extrinsic Factor in Pernicious Anemia

Effect of daily administration of various substances together with 150 cc. of neutralized normal human gastric juice (Castle and associates, 1944)

Substances	Case 99	Case 100
	First Period, 10 days	
"Vitamin free" casein	50 gms. daily	50 gms. daily
all known B vitamins	adequate	adequate
xanthopterin	0.009	0.005
folic acid	0.0036	0.0023
Initial R.B.C. (10^6/cu. mm.)	1.17	1.50
Reticulocyte peak	1.9	1.0
Interpretation	negative	negative
	Second Period, 10 days	
Beef muscle	200	—
Meat extract	—	35 cc.
Initial R.B.C.	1.54	1.79
Reticulocyte peak	13.6	8.9
Interpretation	positive	positive

plement for case 99 and a meat extract for case 100, a positive response resulted as evidenced by a rise in the red blood cell count and the reticulocyte peak. It was concluded that the extrinsic factor for pernicious anemia was not identical with folic acid, xanthopterin, or any of the known B vitamins.

Much effort has been devoted to the isolation of the antipernicious anemia factor in liver. The factor may be multiple. The properties of the concentrates of this factor as studied by five groups of investigators are given in Table XI. The properties of the concentrates differ considerably. For example, the carbon content varied from 41 to 53%, and the nitrogen content varied from 10 to 15%. The molecular weight of one preparation was believed to be in the range of 2000-5000. Two of the materials were reported to have amino nitrogen, but none was found in the material of another group. The fraction of Dakin and his asso-

TABLE XI

Properties of Certain Liver Fractions Containing Antipernicious Anemia Factor

	Cohn, *et al.*	Dakin, *et al.*	Laland, *et al.*	Subbarow, *et al.*	Karrer, *et al.*
Composition					
%C	—	46.8–48.1	53.64	41.56	45.68
%H	—	6.6–6.8	6.85	6.74	6.75
%N	10.8	15.2–16.8	13.33	13.13	14.63
%S	none	—	0.74	1.2	present
Ash	—	—	2.05	—	—
Molecular weight	—	2000–5000	—	—	—
Presence of pentose	none	—	present	present	none
Amino nitrogen					
before hydrolysis	none	0.5	—	5	—
after hydrolysis	none	10–10.4	—	75	—
Amino acids isolated		arginine glycine leucine aspartic acid hydroxyproline proline			

ciates yielded arginine, glycine, leucine, aspartic acid, hydroxyproline, and proline on hydrolysis.

Now that the chemistry of pteroylglutamic acid and its related conjugates is known, it is likely that more rapid progress will be made in the isolation of the liver antipernicious anemia factor(s). The unknown factor or factors for pernicious anemia in liver and other sources are possibly the outstanding unidentified factors of today, if importance is judged by the amount of clinical evidence on therapeutic usefulness.

Just as pteroylglutamic acid has antipernicious anemia activity, pteroyltriglutamic acid has been reported recently by Spies to have activity also. It is effective (Table XII), but pteroylheptaglutamic acid (vitamin Bc conjugate) has been found by Welch and his associates to be ineffective in pernicious anemia in relapse. A possible relationship between the utilization of pteroylglutamic acid and the extrinsic factor is a subject for interesting speculation and study.

About a year ago Cline, Berry, and Spies announced the isolation of a new leukopoietic factor from a liver extract which was active in leukopenia.

TABLE XII

Activity of Folic Acid Conjugates in Pernicious Anemia

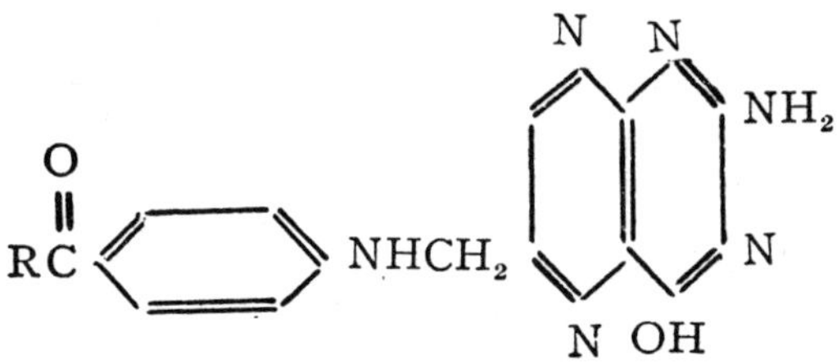

Effective
(Spies, 1946)

R = three glutamic acid moieties
Pteroyltriglutamic acid
(Fermentation *L. casei* factor)

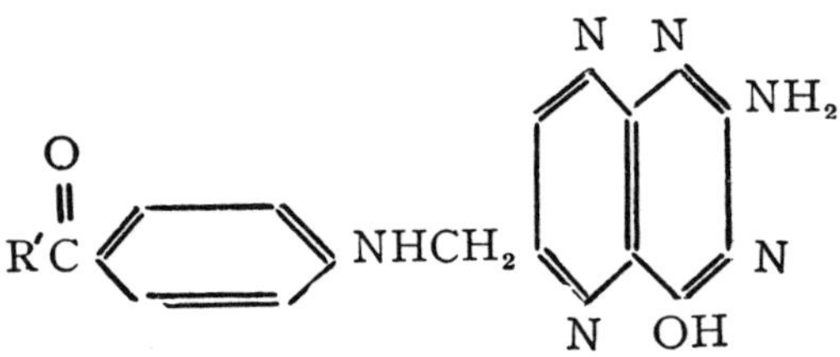

Ineffective
(Welch and associates, 1946)

R' = seven glutamic acid moieties

Pteroylheptaglutamic acid
(Vitamin Bc conjugate)

The liver extract was fractionated with acetone and passed over acid-washed permutit and fractionated again with water-acetone mixtures (Table XIII). The crystals which were obtained appeared to be neutral in character and showed a positive test for primary aliphatic amino groups. The material was active in doses of 20-60 mg. for causing an increase in the white cell count and neutrophiles when given intravenously to malnourished patients with leukopenia. These investigators believed that this product might be related to the extrinsic factor for pernicious anemia.

During 1936-1941 Wulzen and Bahrs studied the growth and disease of planarian worms when the mashed tissues of guinea pigs which were maintained on experimental diets were fed to these worms. The growth of the worms was a measure of a change in the mammalian

TABLE XIII

Isolation of a New Leukopoietic Factor from Liver (Cline, Berry, and Spies, 1945)

Liver extract (reticulogen) —acetone fractionation→ —acid-washed permutit→ —acetone water fractionation→ crystals

Properties of Crystals

1. positive test primary aliphatic amino group
2. contained "ash"
3. 20-60 mg. dose for malnourished patients with leukopenia caused increase in white cell count and neutrophiles

---

tissue which resulted from changing conditions in the animal as a whole. These investigations resulted in the discovery that raw cream contains a factor essential for the guinea pig which differs from those factors previously described for the guinea pig. Deficiency of this factor caused a stiffness of the wrists and elbows of the animals and calcium triphosphate depositions throughout the whole body (Table XIV).

TABLE XIV

A Dietary Factor Essential for Guinea Pigs, Antistiffness Factor

---

Growth and disease of planarian worms fed tissues of guinea pigs (Wulzen and Bahrs, 1936-1941).

Raw cream contains factor for guinea pigs.

Deficiency characterized by stiffness in wrists and elbows and calcium triphosphate depositions.

Partial isolation of factor from raw cream (van Wagtendonk and Wulzen, 1943).

A report on the partial isolation of this factor from raw cream was made in 1943 by van Wagtendonk and Wulzen.

It was found that a deficiency of the antistiffness factor had no effect upon the creatine excretion. In contrast with this result, vitamin E deficiency produced muscular dystrophy accompanied by creatinurea.

The concentration of the easily hydrolyzable phosphorus fraction (including adenosine triphosphate) decreased and the inorganic phosphorus increased in the liver and kidneys during the development of the deficiency. Changes in the adenosine phosphates may involve the metabolism of the purines.

Deficiency of the factor resulted in a lowering of the serum phosphatase level.

Administration of the factor resulted in a return to normal levels for blood phosphorus and calcium, and prevention of the deposition of colloidal calcium phosphate in the muscle. Prolonged administration of the factor resulted in removal of even large deposits of calcium phosphate from the tissues.

Significant changes in the albumin-globulin ratio of the plasma proteins also took place during the deficiency.

It seems apparent that this factor has a regulatory effect upon phosphorus metabolism (Table XV).

The work on the isolation of the antistiffness factor showed that

TABLE XV

Physiological Studies on the Antistiffness Factor (van Wagtendonk and coworkers, 1943-1945)

No effect on creatine excretion—vitamin E deficiency produces severe creatinuria.
Easily hydrolyzable P fraction decreases and inorganic P fraction increases.
Low serum phosphatase level.
Administration of factor resulted in normal levels of blood P and Ca.
Administration of factor resulted in removal of calcium phosphate deposits.
Changes in albumin-globulin ratios in plasma proteins take place.
Conclusion: Factor has regulatory effect on phosphorus metabolism.

raw cream was not a satisfactory source for the successful isolation of the factor since such large amounts of raw cream would be needed. Fifty-five gallons of raw cream gave only 3 mg. of a concentrate of the factor which was curative at a dosage of 0.1γ.

Other raw materials were tested as a source of the factor. Table XVI contains a summary of some of the results of tests on other materials. Potato, beef muscle, liver, and alfalfa have about one-half as much activity as raw cream. Yeast, broccoli, soybean, and beet molasses had no

TABLE XVI

Sources of the Antistiffness Factor

55 gal. raw cream ⟶ 3 mg. concentrate, curative at 0.1γ dosage

Activity of Source Materials (van Wagtendonk and Wulzen, 1946)

Material	*Activity, units/mg.*
Raw cream	1
Raw potato	0.5
Baker's yeast	0
Beef muscle and liver	0.5
Alfalfa	0.5
Broccoli	0
Soy bean	0
Beet molasses	0
Cane molasses	10
Cane juice	100-1000

activity at the levels tested. Cane molasses and cane juice were discovered to be excellent sources of the antistiffness factor and isolation work was started with these raw materials.

A crystalline compound was eventually isolated from crude cane juice. The process for the isolation of the crystalline compound is shown schematically in Table XVII. The crude cane juice showed an activity of 100 u./g. It was first extracted with ether. Distillation of the ether left a residue which showed one million u./g. Distribution of

TABLE XVII

Isolation of the Antistiffness Factor from Cane Juice (van Wagtendonk and Wulzen, 1946)

Crude cane juice (55 gal.) (100 u./g.) —ether extraction⟶ {extract, residual solution} —distillation⟶ residue (10^6 u./g.) —distribution, pet. ether and 90% menthanol⟶

activity in pet. ether solution —distillation⟶ residue (1.5×10^6 u./g.) —MgO, pet. ether and benzene⟶ MgO adsorbate solution —distillation⟶

residue (10×10^6 u./g.) —molecular sublimations and recrystallizations⟶ crystalline material
m.p. 81.5-82°
activity, 500×10^6 u./g.
yield, 100 mg.
dosage, 0.002γ

this residue between petroleum ether and 90% methanol showed that the activity remained in the petroleum ether solution. Distillation of the petroleum ether phase left a residue showing one and one-half million u./g. When this residue was dissolved in a petroleum ether-benzene mixture and treated with magnesia, it was found that many impurities were selectively adsorbed. Removal of the solvent left a residue showing ten million u./g., and two molecular sublimations and many recrystallizations of this material gave a crystalline product which melted constantly at 81.5-82°. It was effective in the biological test at a dosage of 0.002γ and showed 500 million u./g. One hundred milligrams of this crystalline compound was obtained from 55 gallons of crude cane juice.

We may look forward to learning more about the chemistry of this antistiffness factor.

Woolley demonstrated in 1941 the existence of a new growth factor for the multiplication of certain organisms of Lancefield's group A hemolytic streptococci (strains X40, S43, C203 and 594) on a chemically defined medium. It appeared that this factor was required for the growth of other species of bacteria such as pneumococcus D39R. The rate of growth of yeast on the Eastcott medium was also markedly increased (Table XVIII).

TABLE XVIII

A New Growth Factor Required by Certain Hemolytic Streptococci

Growth factor (Woolley, 1941) for: a) Lancefield's group A hemolytic streptococci strains X40, S43, C203 and 594
b) pneumococcus D39R
c) yeast on Eastcott medium

Sources: Liver "Fraction A," yeast and rice bran extracts

Dialyzability: Liver "Fraction A" purification steps → concentrates (dialyzable form)

(non-dialyzable form) — $C_2H_5OH—H_2O—HCl$ → material (dialyzable form)

"Maximal unit": Material/cc. to produce maximal growth and acidity

Purification: Concentrates showed 1 m.u./10γ

Properties: Not soluble in acetone, chloroform or ethyl acetate
Not readily adsorbed by norite (non-dialyzable form)

That fraction of aqueous liver extract which was insoluble in 70% alcohol ("Fraction A"), yeast extract (Difco), and rice bran extract (Vitab) were good sources of the active factor. It appeared that this factor was not a known amino acid, since the organisms did not grow even when the amount of casein hydrolyzate in the medium was increased to one per cent.

It was found that the factor in liver "Fraction A" was not readily dialyzable, but some of the purified concentrates contained the factor in dialyzable form. Treatment of the liver "fraction A" with alcoholic hydrochloric acid converted the non-dialyzable form of the factor into a dialyzable form.

A "maximal unit" for the assay of this factor was defined as the amount of material per cubic centimeter of the medium which was required to produce the maximal growth and acidity.

The purification procedures yielded concentrates which showed 1 m.u./10γ. The factor was not soluble in acetone, chloroform, or ethyl acetate, and was not readily adsorbed on norite.

Pollack and Lindner showed that a substance was present in peptone which was necessary for the growth of *L. casei*, and Smith showed that certain strains of *S. lactis* grew better with a substance present in yeast and peptone. It appeared to Sprince and Woolley that their factor, strepogenin, was probably responsible for the growth phenomena observed by Pollack and Lindner for *L. casei* and by Smith for *S. lactis*. Sprince and Woolley prepared three concentrates of strepogenin by dissimilar methods (dialysis, alcoholic butylamine extraction, and partial hydrolysis of casein, etc.). Each of the three concentrates was assayed for per cent recovery of activity by testing it for the growth of the hemolytic *Streptococcus* X40, *L. casei* and *S. lactis*. It may be seen from these per cent recoveries of activity that the active substance was recovered about equally for these three organisms. Sprince and Woolley concluded that the same substance, or very closely related substances, were responsible for the growth activity in each case (Table XIX).

Since Wright and Skeggs had found strepogenin-like activity for enzymic digests of casein, Sprince and Woolley treated purified proteins with trypsin and examined the digests for strepogenin activity by assays with *L. casei* (Table XX).

The relative strepogenin contents of several such proteins showed that many of them are much richer sources of this growth factor than is the liver fraction L. Crystalline trypsinogen and insulin were outstanding in having 30-40 times as much strepogenin as liver fraction L.

TABLE XIX

Relationship of the Growth Factor (Strepogenin) Required by Certain Hemolytic Streptococci to Growth Phenomena of Other Bacteria

Growth factor (Pollack and Lindner, 1943) for *L. casei*. Source: Peptone
Growth factor (Smith, 1943) for strains of *S. lactis*. Source: Peptone, yeast
Growth factors for *L. arabinosus*, diphtheria bacillus and propionic acid bacteria (1943).

Relative Potencies of Strepogenin Concentrates for Bacteria (Sprince and Woolley, 1944)

Fraction prepared by	Recovery activity (%)		
	strep. X40	*L. casei*	*S. lactis*
Dialysis	54	54	50
Alcoholic butylamine extraction	25	27	21
Partial hydrolysis of casein, etc.	17	19	18

It appeared that the chemical nature of strepogenin is that of an amphoteric peptide-like compound, or it is that of a proper combination of amino acids.

In an effort to learn whether strepogenin was of importance in animal nutrition, Woolley tested the growth response of mice to certain diets.

TABLE XX

The Occurrence of Strepogenin in Purified Proteins

Wright and Skeggs (1944): Casein ⟶ strepogenin-like activity
Chemical nature of strepogenin: amphoteric peptide-like compound or proper combination of amino acids

Relative strepogenin contents of proteinaceous materials (after trypsin digestion for 20 hours), Sprince and Woolley (1945)

Material	*Relative strepogenin content*
Liver fraction L	1
Crystalline insulin	40
Crystalline trypsinogen	30
Crystalline chymotrypsin	16
Crystalline tobacco mosaic virus, etc.	6

The test animals grew at a suboptimal rate or gained only 1.7 grams/week. The addition of 2% casein or 0.5% crystalline trypsinogen to the basal ration caused restoration of the normal growth or a gain of 3.1-3.6 grams/week. The addition of a strepogenin concentrate equivalent to 2% casein also caused normal growth, but the addition of hydrolyzed strepogenin concentrate at twice the casein level (4%) had no effect on the suboptimal growth rate (Table XXI).

These observations and others led to the conclusion that the growth

TABLE XXI

Growth Stimulant in Proteins for Animal Nutrition
Growth Response of Mice to Certain Proteinaceous Materials
(Woolley, 1945)

Addition to basal ration	%	Number of Animals	Average change in weight grams/week
—	—	12	1.7
Casein (vitamin-free)	2	4	3.1
Strepogenin concentrate	eq. 2% casein	8	3.0
Denatured crystalline trypsinogen	0.5	4	3.6
Hydrolyzed strepogenin concentrate	eq. 4% casein	4	1.3

promoting powers of some proteins when added to diets based on amino acids may be attributed to the presence of strepogenin.

Womack and Rose have described the results of their experiments which indicate that proteins contain an unidentified growth stimulant for rats, which is necessary for the maximum increases in weight and which may be identical to strepogenin. For example, they found that rats on a certain amino acid diet gained about 97 grams in weight in 28 days, but when a supplement of 5% casein was added, the average gain was 126 grams. The addition of 5% acid-hydrolyzed casein was considerably less effective in promoting growth (Table XXII). Such experiments indicated that this factor was present in more than seventeen proteins and protein-rich foods.

TABLE XXII

Unidentified Growth Stimulant in Proteins for Rats
Growth response of rats to casein preparations (Womack and Rose, 1946)

Supplement	%	Number of animals	Average total gain in 28 days, grams
—	—	152	97 ± 0.7
Casein	5	10	126 ± 2.5
Acid-hydrolyzed casein	5	7	105 ± 2.9

There are numerous other investigations in progress on other factors. A partial list of these other unidentified factors and vitamins is given in Table XXIII. No doubt some of these nutritional studies are the frontier investigations on new factors which will gain importance in the future.

TABLE XXIII

Partial List of Other Unidentified Factors

Growth factor for *Clostridium sporogenes*	Knight and Fildes (1933)
Vitamin P (permeability vitamin)	Rusznyak and Szent-Gyorgyi (1936)
Factor for reproduction in chickens	McGinnis, Heuser, and Norris (1944)
Feeding factor for chicks	Cravens, McGibbon, and Halpren (1944)
Anemia and weight factors for the pigeon	Street (1944)
Chromotrichia factor(s) for the dog	Frost and Dann (1944)
Unrecognized vitamins for the pig	McRoberts and Hogan (1944)
Liver factor(s) for sulfapyridine anorexia in rats and dogs	Handler (1944)
Factor for growth of *Tetrahymena geleii*	Kidder and Dewey (1944)
Growth factor for *Clostridium perfringens*	Ballentine and associates (1944)
Growth factor for guinea pigs	Woolley and Sprince (1945)
Unidentified factor essential for rat growth	Hartman (1946)
Chick growth factor occurring in cow manure	Rubin and Bird (1946)
Growth factor for *Lactobacillus gayoni* 8289	Cheldelin and Riggs (1946)

V. THE KINETICS OF GROWTH OF MICROORGANISMS

BY C. B. VAN NIEL[1]

THE consensus of opinion, expressed in the preceding chapters, seems to have been that "growth is a very complex phenomenon." I see no reason for amending this verdict when considering the growth of microorganisms. It is true that these creatures appear, even when studied with the best optical or electronic equipment, as rather simple structures; and, compared with higher plants and animals, they certainly do not seem to be as highly differentiated. Nevertheless, they display an astonishing diversity of functions which in many respects they share with the trees, oysters, butterflies, and elephants.

Generally they are found swimming or floating about in an environment containing a more or less abundant variety of chemical compounds. From these they pick up or absorb special ones which as a result of metabolic activities are subsequently converted into cellular constituents. When a large enough quantity and variety of the latter have been produced, and when the individual organism has sufficiently increased in size, it divides into halves, each one of which then repeats the process. This continues until the food supply, or specific ingredients thereof, gives out, or until inhibitory metabolic products begin to interfere with normal development.

The amazing feature of growth, in a microbial culture as in living beings in general, lies in the fact that it involves the manufacture and orderly arrangement of the manifold structural and functional components which characterize the organism in such a way as to duplicate accurately an existing pattern. How this is accomplished is one of the major biological problems, fundamentally as complex in the lower as in the higher forms of life. On the other hand, the comparative ease of handling microorganisms, their rapid growth, and the opportunity for rigorously controlling various environmental factors all combine to render them favorable material for experimental studies of the basic mechanism of the growth process.

In several instances the successful resolution of a complex phenomenon into its constituent components, the latter more readily amenable to scientific analysis than the overall event, has resulted from kinetic

[1] Hopkins Marine Station of Stanford University, Pacific Grove, Calif.

studies. By relating the rate of the process to a number of variables, preferably one at a time, regularities and irregularities are often disclosed which serve as useful indicators of component steps. It therefore seems appropriate to examine what kinetic studies on the growth of microorganisms have contributed and may contribute to a deeper penetration into the problem of growth in general.

Nearly all that is known about the kinetics of growth of microorganisms has been learned from studies of so-called growth curves. By inoculating a specified culture medium with a suitable organism and determining, at the start and after various time intervals, the number of individuals present—frequently restricted to a determination of the number of viable cells—the requisite data are collected for constructing a graph which represents the relationship between the microbial population (or density) and the time elapsed since the initiation of the culture. The most convenient curves are those in which the number of individuals per unit volume is plotted on a logarithmic scale against time. Figure 1 demonstrates the commonly encountered shape of such growth curves.

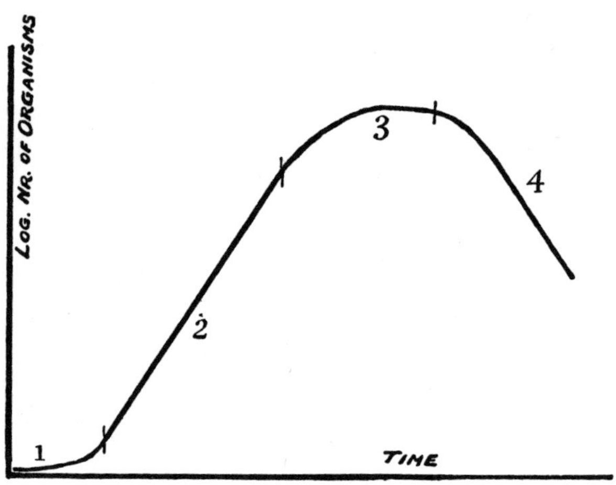

Figure 1. A representative growth curve.

Much attention has in the past been paid to an interpretation of the various phases of these curves. It will be clear that the first phase [1] denotes a period in which a gradual transition takes place from a condition where an increase in numbers does not occur to one where the increase is exponential [phase 2]. This second phase is easily accounted for; it means that during this period all organisms divide regularly, with the daughter cells behaving in an identical fashion. Towards the end of this phase the slope changes back to zero, evidence that when this point is attained [phase 3] either no further divisions take place, or that the increase in cell numbers is exactly balanced by the disappearance or death of individuals. The transition period, often brief and rather sharp, would imply that:

(*a*) either fewer and fewer cells are still in a position to divide, or

(*b*) that the division rate of the individuals becomes progressively less.

During the final phase [4] the death of the cells is predominant; from the point of view of studies on growth this phase is of no interest.

A considerable amount of time has been devoted to the first and originally quite enigmatic phase, referred to as the latency and lag period. As a result of various investigations, mostly before 1930, the attitude towards the problems it presents can be summed up as follows.

In many cases the cause for the existence of a latency and lag period resides in the cells used as inoculum. It can be completely eliminated by inoculating from a culture which itself is in the exponential phase. This has been demonstrated by Chesney (1) for bacterial, and by Phelps (2) for protozoan cultures. The careful observations of Henrici (3) have shown that the morphology of a bacterial species differs markedly according to the phase of the culture. It is therefore clear that the initial lag must be ascribed to a transformation of aged cells into "embryonic forms" (3, p. 140). Depending upon the age of the culture from which the inoculum is prepared, this change may require anywhere from minutes to many hours.

Similarly, a period of preliminary adjustment generally follows upon a change of environmental conditions, such as composition of the culture medium, change in temperature of incubation, etc. In that event a lag phase can be observed even if the new cultures are inoculated from a culture in the exponential phase, and the length of the lag depends upon the extent and nature of the change in conditions.

Growth curves of microorganisms are as a rule accurately reproducible. Chiefly responsible for this feature are (1) the rigorous control which the experimenter can exercise over the environmental conditions, and (2) the small size of the organisms. The small size eliminates many complications connected with the accessibility of foodstuffs. In higher plants and animals transport of nutrients, often over considerable distances, may be a controlling factor for growth; in microorganisms this is not the case. Furthermore, the small size insures that large numbers of individuals are always involved. Hence a statistical treatment of results becomes superfluous; every experiment automatically yields data in which individual variations are represented. This will be obvious if it be realized that with bacterial cultures populations of about 10^9 organisms per ml. are the rule, and that even for pure cultures of protozoa numbers up to 10^6 per ml. are far from rare.

That not all individuals in a culture behave in an identical manner, and that the results obtained really express statistical averages, follow from the fact that growth curves of microorganisms are always smooth even if cultures are started with a single individual, where one might expect a discontinuity such as was first observed by Ellis and Delbrück (4) in their studies on bacteriophage multiplication. More direct evidence for the existence of variation among individuals in bacterial cultures can be found in the investigations of Vaas (5) and Doudoroff (6).

In spite of these advantages, the kinetic approach has not yet been used very effectively or extensively. It is probable that this must chiefly be ascribed to the unwieldy and laborious methodology that used to be necessary for the determination of a growth curve, requiring either many sets of culture plates, or the actual counting under the microscope of thousands of cells for each point on the curve. The newer methodology of determining population densities by turbidity measurements with the aid of photoelectric apparatus, though not always entirely free from objections, may well bring about a renewed interest in this line of work, as is foreshadowed by the important studies of Monod (7) during the last few years.

Kinetic investigations on cultures of microorganisms are eminently suited for establishing relations between growth and environmental factors, especially the nature and amount of nutrients. Based upon the concept developed by Liebig about a century ago, it could be anticipated that in microbial cultures growth would be limited by those ingredients which are present in "minimum" quantity. Experiments such as those of Meyer (8) on the growth of *Aspergillus niger* in media with graded concentrations of phosphate and ammonium sulfate have gone far towards establishing the validity of Liebig's "law of the minimum" for microorganisms. But these were not, strictly speaking, kinetic studies; the measurements dealt with total crops as dependent upon nutrient concentrations rather than with rates of growth. Furthermore, these researches were carried out at a time when little was known concerning the exact mineral requirements of the organism used, which makes the interpretation of the results somewhat difficult.

Decidedly more conclusive in this respect have been the recent studies in connection with growth factor requirements of microorganisms. The mere fact that microbiological assay methods have been developed for many of the known vitamins and amino acids provides convincing evidence for the thesis that a single nutrient factor may determine the extent of growth. However, with a very few exceptions, these contributions too are based upon measuring final yields rather than rates. And where the latter are employed, as for example in the thiamin assay by means of yeast fermentation (9), it is the metabolic, not the growth rates, that are the basis for the determination.

Before 1942 only a few investigations on the kinetics of microbial growth in relation to nutrient concentration were reported. The results have been well summarized by Rahn (10, p. 251) in the following statement:

"With increasing amounts of food, the crop, i.e. the final number of cells in a culture, usually increases.

"With a liberal supply of building material, increase of energy food will cause no acceleration in growth, but a longer growth period, and, therefore, a larger crop. This is limited by accumulation of fermentation products or of inhibiting cell secretions.

"With all food components increasing, the crop will still be limited by fermentation products or cell excretions. The law of diminishing returns becomes quite evident."

While these conclusions were based upon experiments with bacteria and yeasts, they apply equally to the growth of protozoa in pure culture, as was first established by Phelps (11). His studies also revealed a direct proportionality between cell yield and food concentration within very wide limits, a situation which, wrote Phelps, "has not yet been demonstrated in the bacteria nor in most of the work on yeasts." (11, p. 494).

Two years later, however, Dagley and Hinshelwood (12) supplied the experimental evidence that in bacterial cultures also the rate of growth is independent of the concentration of nutrient materials over a wide range, while the final crop of organisms is directly proportional to this concentration. The latter part of these conclusions was corroborated by studies on the growth of *Escherichia coli* in beef broth (13).

All of the earlier studies on the kinetics of growth in relation to nutrient concentration were carried out with complex media in which not one but all constituents were present in varying amounts. It was only recently that Monod (7) started his investigations in a manner more likely to furnish basic information. By using bacteria which can develop in a strictly defined medium, it was possible to determine the effect of variations in the concentration of a single nutrient constituent upon both the rate of growth and the final crop. So far, this approach has been restricted to experiments with sugars and related compounds, applied over a wide range of concentrations. With three different bacterial species it was established that, for a given carbon source, the rate of growth is independent of the substrate concentration, and the total yield is strictly proportional to it, down to such low levels as 20 mg. substrate per liter.

There are on record a number of observations which suggest a different response of microorganisms to the concentration of nutrients. Thorne (14), for example, claims to have shown that the rate of yeast growth in sugar media falls uniformly with increasing concentration of the yeast, a behavior entirely at variance with the above-mentioned results. And Porter (15) has recently resuscitated the concept of "biological space" as earlier developed by Bail (16). The argument here is that, because the total crop of bacteria in liquid cultures reaches a certain maximum regardless of the concentration of foodstuffs, growth must be limited by a space factor rather than by nutrients. The following passage from Porter's book (15, p. 136) is worth quoting in this connection:

"There is considerable evidence in favor of some of Bail's claims concerning space theory and *M*(maximum)-concentration, but many of his points are not accepted at this time. Further work will have to be done on this subject—why bacteria stop dividing at a maximum speed when a certain population is reached—before any definite conclusions can be reached."

Most, if not all, of these aberrant results can be readily ascribed to the experimental conditions imposing a limiting factor which, in spite of repeated exhortations, is still insufficiently taken into account. This limiting factor is oxygen. The supply of this gas is governed by its very low solubility coefficient and unless special measures are adopted to insure a greatly increased rate of diffusion (e.g. by forced aeration, continuous shaking, etc.), its availability in liquid media is apt to fall below that required for an optimum growth rate of the organisms. The studies of Phelps (11), Rottier (17), Rahn and Richardson (18), and Anderson (19) attest to the correctness of this statement. It seems more than likely that the "biological space" referred to above is simply a measure of the oxygen supply. Particularly the close packing of bacteria in colonies provides a convincing argument for the contention that space *per se* cannot be considered seriously as an important factor in growth.

Thus the best accessible evidence indicates that the growth rate is not affected by the concentration of nutrients between very wide limits, and that the total yield, over an equally wide range, is proportional to the amount of food. This situation carries important implications which will next be considered.

Let me reiterate that growth consists in part of the transformation of foodstuffs into cellular constituents. This, in turn, depends upon metabolism. In Chapter III Thimann has presented a number of cases in which growth responses are strictly paralleled by metabolic responses following upon the exposure of plants to auxin solutions. It is therefore of the greatest significance that the rate of metabolism, as has been repeatedly ascertained, is generally unaffected by concentration differences of the substrate down to extremely low levels.[2] And it is not surprising that growth would follow the same pattern.

In many respects the relation between metabolism and growth is quite obvious. "No growth without metabolism" was a dictum as familiar since the latter part of the nineteenth century as the phrase "No thought without phosphorus" had been earlier. The reason for this has become apparent as the concepts of thermodynamics were developed. The growing cell must synthesize cellular constituents, and for these syntheses energy is required.

As a first approximation the relation between growth and metabolism could thus be understood as one of energy linkage; the metabolic activities of the cell provide the energy necessary for the performance of synthetic reactions. Based upon this concept, various investigations have been conducted aimed at measuring the "energetic yield" of growth processes. By determinations of the amount of cell material that could be synthesized from a known quantity of substrate, of the caloric value of the cell material, and of the energy liberated by oxidations accompanying the observed growth, data were assembled which permitted the computation of such "energetic yields." In subsequent studies of the same general nature more precise relations were calculated by distinguishing between a "metabolism for upkeep" and a "metabolism for growth" (20, 21).

However useful and important this sort of information may have been, recent trends have brought about a considerable change in the

[2] Some interesting cases have been found in which this is not true. These, however, present problems which are outside the scope of this discussion.

primary approach to the problems of syntheses and growth. Biochemists in particular have come to recognize that the thermodynamic formulations beg the question which has become paramount in today's biochemistry, viz. that of the mechanism whereby the result is achieved. Applied to the problem under discussion, this means that the concept of an energetic coupling between synthesis and breakdown, no matter how sound fundamentally, is too vague to be permanently satisfactory, and that ultimately our task is to discover the mechanisms whereby such coupling is accomplished.

The most fruitful hypothesis that has been advanced in this connection assumes the existence of material links between breakdown and synthesis. It may, somewhat primitively perhaps, be paraphrased in this manner: instead of regarding biological syntheses as very special processes which, because they require energy, are basically different from the breakdown reactions in which energy is liberated, it would be more appropriate to describe syntheses as series of enzyme-controlled step reactions, each step comparable in every way but one with those which collectively compose the breakdown. The differentiating feature is that, whereas the breakdown reactions can always start with any one of the utilizable substrates in the food, the synthetic processes require particular substances which are not usually provided but which arise from the initial substrates as intermediate products in the normal course of their metabolic degradation. Such products are characterized by being more unstable, i.e. more reactive, than their ultimate precursors, the foodstuffs proper, thus implying that their energy level has been raised with respect to the initial materials.

This concept, first enunciated by Kluyver (22, 23) some fifteen years ago, does not conflict with the earlier one of energetic coupling. Its great value lies in the penetrating insistence upon the functioning of an intelligible chemical mechanism which is capable of accounting for the transfer of energy.

During the past few years tremendous strides have been made in this direction. Ranking high among these is the discovery of the phosphorylated carbohydrate now known as "Cori-ester," a compound which can serve as the immediate raw material for the spontaneous enzymatic synthesis of a number of di- and polysaccharides (24-30). No less significant are the studies of Lipmann (31, 32) on the labile intermediary metabolic products characterized by a high-energy phosphate bond. It has now been shown that these substances, too, can be used for performing a number of fundamental biochemical syntheses with isolated enzyme systems.

A special corollary of Kluyver's concept is that the molecular structure of the substrate and of the intermediate products to which this can lead during its breakdown, rather than the amount of energy liberated during this process, becomes the all important determinant of its potentialities as the starting point for the synthesis of cell materials. The investigations by Clifton and Logan, Doudoroff, and Bernstein on the assimilation of chemically related compounds at different energy levels (33-36) provided suggestive results in this connection. Better yet can this situation be illustrated by recent studies on the synthesis of some characteristic cell constituents.

Having recognized a number of vitamins as parts of enzyme systems, it has become increasingly clear that the biosynthesis of such entities proceeds in a chemically intelligible manner, even though many of the steps involved in their manufacture may still be obscure. The mere fact that certain microorganisms are unable to develop in simple media without being supplied with preformed parts of enzymes shows conclusively that the supply of energy is not the governing factor, and that the availability of appropriate building blocks must be reckoned among the requirements for growth. Such phenomena are now generally interpreted to mean that the cells lack the machinery for the synthesis of essential compounds, which must consequently be supplied. It has frequently been shown that the same organisms can also develop in an environment which contains, instead of the enzyme part itself, substances that are chemically more or less closely related to it, and from which the cellular constituent in question is elaborated.

Corresponding observations have been made pertaining to the synthesis of specific amino acids, as the epoch-making investigations of Beadle and Tatum and their collaborators with "biochemical mutants" have so amply demonstrated. Through studies of this kind it has been possible to elucidate various steps in the biosynthesis of a number of cell materials and to develop the thesis that each step is controlled by a specific gene. It is impossible to do this field justice here; reference is made to some recent reviews (37, 38).

No student of the chemistry and physiology of growth is likely to deny the importance of these advances in our knowledge of biosynthetic mechanisms. Nevertheless the inclusion of this phase may appear as a digression from my topic. What have such problems to do with the kinetics of growth of microorganisms?

It will be evident that even the growth of a microbe is subjected to the "principle of limiting factors." If the synthesis of all cell constituents can proceed faster in one medium than in another, growth will obviously be faster in the former. Similarly, the rate at which a single component is being synthesized may, under special circumstances, be the determining factor for the rate of growth of the organism. In that case our understanding of biosynthetic mechanisms might be advanced by kinetic studies on growth, as well as by the more strictly chemical ones which hitherto have been the sole method applied. And through a kinetic approach it may be possible to learn something about the synthesis of these cell constituents by organisms whose external environment does not need to contain a supply of closely related substances because the biosyntheses can proceed from a variety of simple ingredients of the medium.

While no investigations have yet been conducted with this purpose in mind, some data are available which support the idea here advanced. For example, Monod includes in his monograph (7) figures on the rate of growth of *Bacillus subtilis* in a complex and in a synthetic medium, the latter composed of inorganic substances and sucrose. Growth proceeds faster in the former than in the latter medium; the generation times are 36 and 50 minutes respectively. Such differences could well be due to differences in the rate at which particular cell constituents become available, especially in view of the fact that in the synthetic solution they must all be built up from inorganic ingredients and sucrose. If it were possible to increase the growth rate in the simple medium through the addition of known chemical entities, it might also become clear what special syntheses are responsible for the lower growth rate.

Another instance is provided by Whelton and Doudoroff's (39) study on the assimilation of sugars and related compounds by *Pseudomonas saccharophila.* It includes a determination of the generation times of this bacterium in a standard mineral medium with any one of a number of carbon compounds. These were found to vary from 105-110 minutes in the presence of trehalose and sucrose, to 196 minutes in the presence of acetate; cultures in glucose and lactate exhibited an intermediate behavior with generation times of 178 and 136 minutes respectively. Studies of this kind, if sufficiently extended, could yield information concerning the nature of intermediate products which serve as building blocks and the rate at which these are generated. Correlation of such data with overall growth rates, supplemented with experiments on the influence of special additions selected for the purpose of furnishing similar or different building blocks, might then reveal the nature of limiting synthetic reactions.

I do not mean to convey the impression that the above-suggested approach is an easy and certain way of solving the problem of growth. They are intended to emphasize that kinetic studies of the growth of microorganisms, hitherto rather neglected, should be seriously considered as potentially capable of supplying information which can contribute toward a more detailed analysis of the manifold and interlinking events which, in their totality, constitute the growth process.

The foregoing reflections on the effects of external factors on the growth rate of microorganisms, and on the interpretations and deductions that can be derived from the observed results, may have fostered the impression that internal factors can be regarded as more or less constant. But this is not the case; there exists an impressive body of indisputable evidence to the contrary. Changes in the morphological, physiological, and biochemical behavior of microbial cultures, attributed to adaptations, modifications, mutations, etc., have long been recognized as manifestations of the intrinsic variability of the organisms. Monod's recent studies (7, 40-42) on the growth rate of bacteria in media containing two different carbon sources instead of a single one offer an important advance in this field.

Depending upon the nature of the organism and of the substrates, either of two types of growth curves describes the observed response of the bacteria. One represents a single period of exponential growth extending over the entire time interval during which substrate is available, with the sharp break at the end marking the complete utilization of the two carbon sources. The other kind consists of two distinct parts, each in itself a characteristic growth curve, with its phases of accelerated, maximal, and retarded growth rates, the two curves being conspicuously separated by a horizontal stretch which indicates the cessation of growth during the corresponding period. Monod has designated this latter behavior by the term "phénomène de diauxie" ("diphasic growth") and has shown that it involves the occurrence of an adaptation process.

By chemical analyses of the medium at different times it was ascertained that the first part of the growth curve corresponds to growth of the organism at the expense of only one of the two substrates, to the complete exclusion of the other. The second, separate growth curve coincides with the decomposition of the second substrate. And the intervening period during which the population remains constant represents the interval needed by the bacteria to adapt themselves to the utilization of this second substance.

Adaptations of this sort have been known for several years and have been ascribed to an adaptive enzyme formation (43, 44). They are strictly dependent upon the presence of a specific substrate, an observation which has led Yudkin (45) to develop an ingenious hypothesis concerning the mechanism by which adaptive enzymes are generated, based upon differences in stability of the free enzyme molecules and of the enzyme-substrate complex. Monod's experiments (40) have shown that the rate of such adaptations in bacterial cultures depends not only upon the presence but also upon the concentration of the substrate. In view of the well-established fact that metabolic rates are generally independent of this last-named factor, and also that the growth rate of the adapted bacteria is constant over a wide concentration range, Yudkin's hypothesis may have to be modified.

Later experiments by Monod (42) have made it probable that adaptive enzyme formation must be interpreted as a competitive phenomenon. This follows from observations on diphasic growth in relation to variations in the concentration ratio of the two substrates. The higher this ratio in favor of the substrate inducing the adaptation, the shorter is the period of suspended growth: the measure of the adaptation rate. The lag can even be eliminated completely by sufficiently magnifying the difference in concentrations of the substrates.

From these results it has been inferred that the two different substrates compete for a common precursor of the specific enzymes. Differences in affinity of the substrates for the precursor would then be responsible for the quantitative variations in the enzyme systems of organisms exposed to solutions which contain both substrates simultaneously but in different concentration ratios.

Growth is the expression *par excellence* of the dynamic nature of living organisms. Among the general methods available for the scientific investigation of dynamic phenomena, the most useful ones are those which deal with the kinetic aspects. The relative paucity of such data on the growth of microorganisms at the present time cannot be denied.

Nevertheless a few contributions of considerable importance have been made in this respect. It can also be asserted with some confidence that the recent advances in methodology for the determination of growth curves will help in overcoming the justifiable hesitation on the part of investigators, previously confronted with cumbersome techniques, to adopt this approach for future studies on microbial growth. Only by further experimentation can the potentialities of kinetic studies in this field be duly appreciated and more clearly defined.

REFERENCES

1. Chesney, A. The latent period in the growth of bacteria. *Jour. Exper. Med., 24,* 387-418, 1916.
2. Phelps, A. Growth of protozoa in pure culture. I. Effect upon the growth curve of the age of the inoculum and of the amount of the inoculum. *J. Exp. Zool., 70,* 109-130, 1935.
3. Henrici, A. T. *Morphologic Variation and the Rate of Growth of Bacteria.* Charles C. Thomas, Springfield, Ill., 1928.
4. Ellis, E. L., and M. Delbrück. The growth of bacteriophage. *J. Gen. Physiol., 22,* 365-384, 1939.
5. Vaas, K. F. *Studies on the Growth of Bacillus megatherium de Bary.* Diss. Leiden, 1938.
6. Doudoroff, M. Experiments on the adaptation of Escherichia coli to sodium chloride. *J. Gen. Physiol., 23,* 585-611, 1940.
7. Monod, J. *Recherches sur la croissance des cultures bactériennes.* Hermann & Cie., Paris, 1942.
8. Meyer, R. Zum Ertragsgesetz bei Aspergillus niger. *Arch. f. Mikrobiol., 1,* 277-303, 1930.
9. Schultz, A. S., L. Atkin, and C. N. Frey. Determination of vitamin B_1 by yeast fermentation method. *Ind. Eng. Chem.,* Anal. Ed., *14,* 35-39, 1942.
10. Rahn, O. *Physiology of Bacteria.* Blakiston, Philadelphia, 1932.
11. Phelps. A. Growth of protozoa in pure culture. II. Effect upon the growth curve of different concentrations of nutrient materials. *J. Exp. Zool., 72,* 479-496, 1936.
12. Dagley, S., and C. N. Hinshelwood. Physicochemical aspects of bacterial growth. I. Dependence of growth of Bacterium lactis aerogenes on concentration of medium. *J. Chem. Soc., 1938,* 1930-1936, 1938.
13. Korinek, J. Zum Problem der Bakterienpopulation. *Centr. f. Bakt.,* II. Abt., *100,* 16-25, 1939.
14. Thorne, R. S. W. The nitrogen nutrition of yeast. *Wallerstein Laboratories Communic., 9,* 97-114, 1946.
15. Porter, J. R. *Bacterial Chemistry and Physiology.* Wiley, New York, 1946.
16. Bail, O. O. možnosti a základech pokusné nauky populační. Čas. Čes. lék. 1927; *Immunitäts f., 60,* 1-22, 1929.
17. Rottier, P. B. Recherches sus les courbes de croissance de *Polytoma uvella.* L'influence de l'oxygénation. *C. rend. Soc. Biol., 122,* 65-67, 1936.
18. Rahn, O., and G. L. Richardson. Oxygen demand and oxygen supply. *Jour. Bact., 41,* 225-249, 1940; *ibid., 44,* 321-332, 1942.

19. Anderson, E. H. Nature of the growth factor for the colorless alga Prototheca zopfii. *J. Gen. Physiol., 28*, 287-296, 1945.
20. Algera, L. Energiemessungen bei Aspergillus niger mit Hilfe eines automatischen Mikro-Kompensations-Calorimeters. *Diss. Gronigen*; *Rec. Trav. botan. néerland., 29*, 47-163, 1932.
21. Tamiya, H. *Le Bilan matériel et l'énergétique des synthèses biologiques.* Paris, Hermann & Cie., 1935.
22. Kluyver, A. J. Atmung, Gärung und Synthese in ihrer gegenseitigen Abhängigkeit. *Arch. f. Mikrobiol., 1*, 181-196, 1930.
23. Kluyver, A. J. *The Chemical Activities of Microorganisms.* Univ. of London Press, 1931.
24. Cori, G. T., and C. F. Cori. The kinetics of the enzymatic synthesis of glycogen from glucose-1-phosphate. *J. Biol. Chem., 135*, 733-756, 1940.
25. Cori, C. F. Phosphorylation of glycogen and glucose. In *Biological Symposia, 5*, 131-140. Lancaster, Jacques Cattell Press, 1941.
26. Hanes, C. S. The breakdown and synthesis of starch by an enzyme system from pea seeds. *Proc. Roy. Soc., B, 128*, 421-450, 1940.
27. Hassid, W. Z. The molecular constitution of starch and the mechanism of its formation. *Quart. Rev. Biol., 18*, 311-330, 1943.
28. Doudoroff, M., N. Kaplan, and W. Z. Hassid. Phosphorolysis and synthesis of sucrose with a bacterial preparation. *J. Biol. Chem., 148*, 67-75, 1943.
29. Doudoroff, M., W. Z. Hassid, and H. A. Barker. Synthesis of two new carbohydrates with bacterial phosphorylase. *Science, 100*, 315-316, 1944.
30. Hassid, W. Z., M. Doudoroff, and H. A. Barker. Enzymatically synthesized crystalline sucrose. *J. Amer. Chem. Soc., 66*, 1416-1419, 1944.
31. Lipmann, F. Metabolic generation and utilization of phosphate bond energy. *Advances in Enzymology, 1*, 99-162, 1941.
32. Lipmann, F. Acetyl phosphate. *Advances in Enzymology, 6*, 216-268, 1946.
33. Clifton, C. E., and W. A. Logan. On the relation between assimilation and respiration in suspensions and in cultures of Escherichia coli. *Jour. Bact., 37*, 523-540, 1939.
34. Doudoroff, M. The oxidative assimilation of sugars and related substances by Pseudomonas saccharophila. *Enzymologia, 9*, 59-72, 1940.
35. Bernstein, D. E. Studies on the assimilation of dicarboxylic acids by Pseudomonas saccharophila. *Arch. Biochem., 3*, 445-458, 1944.
36. Clifton, C. E. Microbial assimilations. *Advances in Enzymology, 6*, 269-308, 1946.
37. Knight, B. C. J. G. Growth factors in microbiology. *Vitamins and Hormones, 3*, 105-228, 1945.
38. Beadle, G. W. Biochemical genetics. *Chem. Rev., 37*, 15-96, 1945.

39. Whelton, R., and M. Doudoroff. Assimilation of glucose and related compounds by growing cultures of Pseudomonas saccharophila. *Jour. Bact., 49,* 177-186, 1945.

40. Monod, J. Influence de la concentration des substrats sur la rapidité d'adaptation chez le B. coli. *Ann. Inst. Past., 69,* 179-181, 1943.

41. Monod, J. Remarques sur le problème de la spécificité des enzymes bactériens. *Ibid., 70,* 60-61, 1944.

42. Monod, J. Sur la nature du phenomène de diauxie. *Ibid., 71,* 37-40, 1945.

43. Karström, H. Über die Enzymbildung in Bakterien und über einige physiologische Eigenschaften der untersuchten Bakterienarten. Diss., Helsingfors, 1930.

44. Karström, H. Enzymatische Adaptation bei Mikroorganismen. *Ergebn. Enzymforsch., 7,* 350-376, 1938.

45. Yudkin, J. Enzyme variation in microorganisms. *Biol. Rev., 13,* 93-106, 1939.

VI. CELLULAR METABOLISM AND GROWTH

BY E. S. GUZMAN BARRON[1]

It is generally recognized that the energy necessary for cellular activities, whether it be used to maintain the steady state of adult life, the "dynamic equilibrium" of Schoenheimer, or the state of growth and synthesis which predominates in dividing and growing cells, is provided by enzymatic reactions largely of the oxidation-reduction type. This has been amply demonstrated by the increased O_2 utilization during growth and by the necessity of aerobic processes for the synthesis of carbohydrates, fats, and proteins which together make the bulk of the cells.

It will be therefore necessary before discussing the role of enzyme reactions in cellular growth to review some fundamental aspects of the properties of the oxidation-reduction enzymes.

General Properties of Enzymes

Enzymatic oxidation-reduction reactions are distinguished from chemical reactions in general by the facts that they are reversible, that electron exchange occurs in steps, and that they have a great tendency to form coupled reactions.

The reversibility of enzyme reactions can be accepted as a general property, since more and more of those reactions hitherto considered irreversible are being shown to be reversible. Witness the enzymatic synthesis of phosphopyruvic acid from pyruvic acid (48), a reversal of the dephosphorylation of phosphopyruvic acid, which had been reported as an irreversible process (58). In many instances reversibility of oxidation-reduction reactions is made easy by the entrance of P into the reactions and generation of energy-rich phosphate bonds, as was shown by Warburg and Christian (83) in the compulsory coupling between the oxidation of phosphoglyceraldehyde and the phosphorylation of adenylic acid, a finding extensively generalized to other reactions by Lipmann (53).

The first step in an enzymatic reaction is the formation of the enzyme-substrate complex. Evidence in favor of such a concept, introduced by Michaelis and Menten (61), is abundant. Knowledge of the nature of

[1] The Chemical Division, Department of Medicine of the University of Chicago.

this complex formation awaits clarification of the structure of proteins, since it is probably the protein moiety which combines with the substrate. The substrate to be oxidized must fit quite closely to the side chains of the protein molecule where combination occurs. No lock, however, remains immune to the prying of false keys; and substances with chemical structure close to that of the substrate will compete for union with the protein, thus affecting the activity of the enzyme. This is strikingly demonstrated by the inhibition of acetate oxidation by fluoroacetate discovered in animal tissues (Bartlett, Ahrens, and Barron [14]) and in yeast (Kalnitsky and Barron [42]). In yeast, acetate oxidation, completely inhibited by monofluoroacetate, was not affected at all by the other halogen acids, bromo-, iodo-, chloro- acetic acids, as well as by trifluoroacetic acid (Table I). This inhibition is due to compe-

TABLE I

Effect of Halogen Acetates on the Oxidation of Acetate by Baker's Yeast. *p*H, 6.2. Temp. 25°. Acetate concentration, 0.01 *M*; halogen acetate, 0.001 *M*.

Halogen	Inhibition (%) Acetate Oxidation	Inhibition (%) Endogenous Respiration	Interatomic Distances C—Halogen bond (A)
CH_2FCOOH	95	24	1.41
$CH_2ClCOOH$	none	none	1.76
$CH_2BrCOOH$	4.8	40	1.91
CH_2ICOOH	1.5	none	2.10
CH_3COOH	—	—	1.09
CF_3COOH	none	none	—

tition of fluoroacetate with acetate in the protein of the enzyme, for inhibition occurs in the first step of acetate metabolism, namely citric acid formation (Table II). Of all these acids it is monofluoroacetic acid which approaches most closely in size acetic acid, as the interatomic distance between C and F (1.41 A) is only 0.32 A more than the interatomic distance between C and H. An increase in the interatomic distance of 0.35 A as in C and Cl (1.76 A) was enough to avoid penetration of the halogen acid into the space reserved for acetic acid in the protein moiety. The protein-substrate complex formation resembles thus strikingly the antibody-antigen complex formation in the specificity.

Some proteins require the presence of —SH groups for enzymatic activity. Furthermore, in some cases only a portion of the —SH groups

TABLE II

Effect of Fluoroacetate on Citrate Formation by Baker's Yeast.

Yeast cells, 8.5 mgs.; Mg acetate, 0.077 *M*; fluoroacetate, 0.005 *M*. Duration of experiments, 4 hours. Blank values for oxygen uptake and citric acid have been subtracted.

Experiment No.	Citric acid formation (cmm.)		Inhibition (%)	O_2 Uptake (cmm.)	
	Control	Inhibitor		Control	Inhibitor
I	4.99	0.045	91	1010	0
II	3.86	0.038	90	1005	0
III	5.0	1.4	72	1012	242

present in the protein are necessary for enzyme activity, as demonstrated with urease by Hellerman and coworkers (35). A survey made by Barron and Singer (13, 78) showed that a large number of enzymes which catalyze the metabolism of carbohydrates, fats and protein derivatives need the electronegative —SH groups for activity (Table III). Doubtless there are still other enzymes, not yet studied, which belong to this class. An indication in favor of this assumption is given by the inhibition of mitosis produced by small doses of x-rays and by nitrogen mustards. Up to the time of Dale's experiments (26) little attention was paid to enzyme inhibition by x-rays. In 1940 Dale demonstrated that enzyme inhibition could be produced by small doses of x-rays provided protein concentration was kept low. Barron, Dickman, and Singer (8) have shown that x-rays, α, and β particles and γ radiations inhibit sulfhydryl enzymes at small doses by oxidation of the —SH groups of the protein (Tables IV and V). Inhibitions produced by x-rays and α particles could be released on addition of glutathione. This inhibition is undoubtedly produced by the OH and OH_2 radicals and H_2O_2 formed by the ionization of water during irradiation. In fact, protection of the —SH groups by mercaptide formation with Cl-Hg benzoic acid abol-

TABLE III

Enzymes Requiring —SH Groups for Enzymatic Activity.

I	II	III
1. Phosphoglyceraldehyde dehydrogenase	11. Acetate *oxidase*	26. Papain
2. Carboxylase	12. Stearate dehydrogenase	27. Urease
3. Pyruvate dehydrogenase	13. Oleate dehydrogenase	28. d-Amino acid dehydrogenase
4. Pyruvate dismutation	14. β-Hydroxybutyric dehydrogenase	29. l-Glutamate dehydrogenase
5. Ethanol dehydrogenase (yeast)	15. Lipase	30. Monoamine dehydrogenase
6. Glycerol dehydrogenase (bacteria)	16. Esterase	31. Transaminase
7. Malate dehydrogenase	17. Acetylase	32. Cathepsin
8. α-Ketoglutarate dehydrogenase	18. Choline dehydrogenase	33. Hemolysin (pneumococcus)
9. Succinodehydrogenase	19. Lysozyme	34. Hemolysin (cobra venom)
10. β-Amylase (barley)	Phosphorylations	
	20. Adenosinetriphosphatase (myosin)	
	21. Myokinase	
	22. Adenylpyrophosphatase (liver)	
	23. Hexokinase	
	24. Phosphorylase	
	25. Phosphoglucomutase	

ished inhibition of urease by γ radiation (Table VI). Destruction of H_2O_2 by catalase, diminished the degree of inhibition of phosphoglyceraldehyde dehydrogenase by α particles, and suppressed the inhibition by β particles. The combination of nitrogen mustards with —SH groups

TABLE IV

Effect of X-Rays on the Activity of Phosphoglyceraldehyde Dehydrogenase. Buffer, phosphate, *p*H 7.0. Enzyme activity, $K = \frac{1}{t} \times \frac{Co - C}{CoC}$ where *t*, time in minutes; Co (DPN) initial; C (DPN) at time *t*.

X-Ray dose (r)	Enzyme Activity $K \times 10^5$	Inhibition (%)	Reactivation with Glutathione (0.002 *M*)
none	4.80	—	—
25	4.80	none	—
50	4.80	none	—
100	3.81	21	complete
200	2.43	50	62
300	0.98	80	—
400	0.98	80	—
500	0.25	94	10

TABLE V

The Effect of α (Po), β (Sr^{89}), and γ (Ra) Radiation on the Activity of Phosphoglyceraldehyde Dehydrogenase. Enzyme, 20γ; phosphoglyceraldehyde, 4×10^{-7} *M*; DPN, 2.5×10^{-7} *M*; $NaHASO_4$, 6×10^{-6} *M*; pyrophosphate buffer, *p*H 8.5, 3×10^{-5} *M*. Temp. 26°.

α Radiation		β Radiation		γ Radiation	
Dose (r)	Inhibition (%)	Dose (r)	Inhibition (%)	Dose (r)	Inhibition (%)
180	58	20	10	25	17
360	73	40	17	50	54
520	92	80	32		
1080	complete	120	37	200	67

is known. However, as this combination is irreversible, it is difficult to establish definitely whether enzyme inhibition is due to this reaction or to combination with some other group of the protein. Very little is known about the role of these —SH groups. Protection of the —SH groups of the protein moiety of succinoxidase by succinate and by malonate (37) would suggest that the substrate is anchored in close

TABLE VI

Inhibition of Crystalline Urease by Radium Radiation
Effect of Glutathione, Glycine, and p-Cl-Hg Benzoic Acid.

Dosage (r)	Additions	Inhibitions (%)
100	—	24
100	glutathione	30
200	—	36
100	glycine, 1×10^{-6} *M*	none
100	glycine, 1×10^{-1} *M*	none
200	—	34
200	glutathione	39
200	p-Cl-Hg benzoic acid+glutathione	none

vicinity to the —SH groups. The proximity of the carboxyl electronegative groups would decrease the activity of the —SH groups and thus render them less susceptible to the action of sulfhydryl reagents. Some other proteins require the presence of a metal (Mg, Mn, Zn, Co) for enzyme activity; the metal does not participate in the reaction. Their role is still more obscure than that of the —SH groups. It is known, however, that —SH groups and metals contribute to increase the stability of the native protein.

The combination of the substrate with the protein brings forth the "activation" of the substrate. It is not yet clear what is meant by activation of the substrate. It is plausible, however, to postulate with Michaelis that activation consists in the formation of a radical, the first step in the process of oxidation of organic substances with a loss of two electrons. This is implicit in Michaelis' theory of compulsory univalent oxidation (60).

In condensation reactions where two substrates take part in the reaction, as in the formation of acetylcholine, it seems that only one of the substrates combines with the protein and becomes activated. Evidence for this assumption is provided by inhibitions with competitive inhibitors. For example, methyl bis(dichloroethyl)amine HCl inhibits choline oxidase because of the similarity of its first transformation product, ethylene imonium, to choline; it also inhibits the synthesis of acetylcholine and its hydrolysis. It has no effect on the hydrolysis of tributyrin. Fluoroacetate inhibits the oxidation of acetate while it has no effect on the synthesis of acetylcholine and acetylations in general (Table VII). It must be concluded that in acetylations the protein combines with the acetylated compound and not with the acetyl group.

TABLE VII

Effect of Methyl bis (Dichloroethyl) Amine (0.001 *M*) and Fluoroacetate (0.01 *M*) on Acetylations, Oxidations, and Hydrolysis.

Reaction	Inhibition (%)	
	N Mustard	Fluoroacetate
Acetylation of sulfanylamide	none	none
Acetylation of choline	75	none
Oxidation of choline	complete	none
Oxidation of acetate	none	complete
Choline esterase (hydrolysis of acetylcholine)	complete	—
Esterase (hydrolysis of tributyrine)	none	—

A number of proteins, components of enzyme systems, have been isolated and prepared in crystalline form. Furthermore our knowledge of the physical and chemical properties of proteins has considerably increased during the last years. The time has come to use these techniques for a serious study of the protein moiety of the enzyme. It is necessary to know the architecture of the protein moiety of the enzyme, the spacing of the side groups to which the substrates attach themselves, and the nature of the activation of the substrate.

The Oxidation-Reduction Systems

In 1930 Barron and Hoffman (10) found that the catalytic power of reversible oxidation-reduction dyes on cellular respiration was conditioned by two factors: the oxidation-reduction potential of the dye (Fig. 1), and the permeability of the cell membrane (Table VIII).

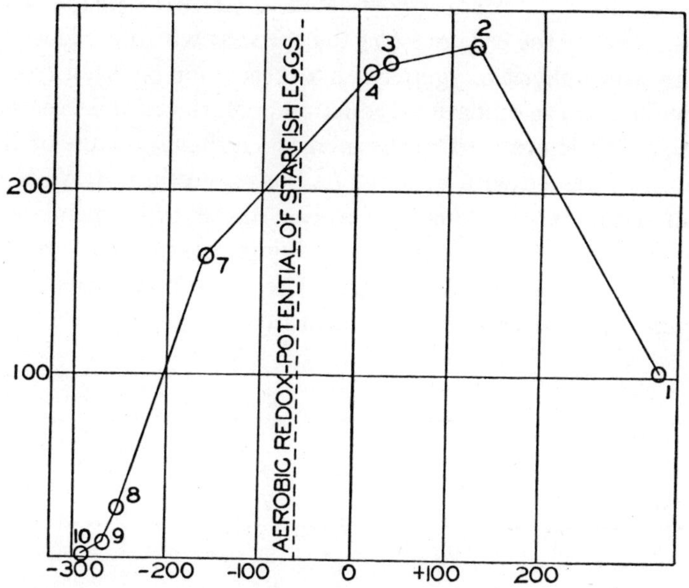

Figure 1. Relation between the catalytic power of dyes and its E_0' to the oxidation-reduction potential of starfish eggs. (Dyes which penetrate through the cell wall.) Abscissa, E_0' (in millivolts) of dyes at *p*H 7.0. Ordinate, per cent increase of O_2 consumption after dye addition. 1. phenol-indophenol; 2. toluylene blue chloride; 3. methylene blue; 4. cresyl blue; 7. cresyl violet; 8. safranine; 9. janus green; 10. neutral red. From Barron and Hoffman, *J. Gen. Physiol.*, *13*, 483, 1930.

TABLE VIII

The Effect of Reversible Dyes on the Oxygen Consumption of Starfish Eggs (Mature, Unfertilized). The Effect of the Membrane Permeability.

Name of dye	E_0' at *p*H 7.00 in volts	Oxygen consumption per hour in cmm.		Permeability of membrane	Per cent increase
		Before dye addition	After dye addition		
Gallocyanine	+0.013	23.1	33.7	does not penetrate	46
Methylene blue	+0.011	23.3	85.5	penetrates	267
Indigodisulphonate	−0.125	25.4	24.6	does not penetrate	none
Cresyl violet	−0.165	25.3	65.8	penetrates	160

The speed of this catalysis was correlated to (*a*) the speed at which the dye was reduced by the cell, and (*b*) the speed at which the leucodye was oxidized by atmospheric oxygen. Such a correlation between free energy (as determined by the oxidation-reduction potential of the dye) and rate of oxidation (as determined by the increase in the O_2 uptake of the cell) was naturally received with caution. At the suggestion of W. Mansfield Clark and with his generous cooperation these experiments were repeated in homogenous systems by studying the rate of oxidation of reduced dyes by atmospheric oxygen. The same relation between free energies and rates of reaction was found (4) (Fig. 2). The theory of

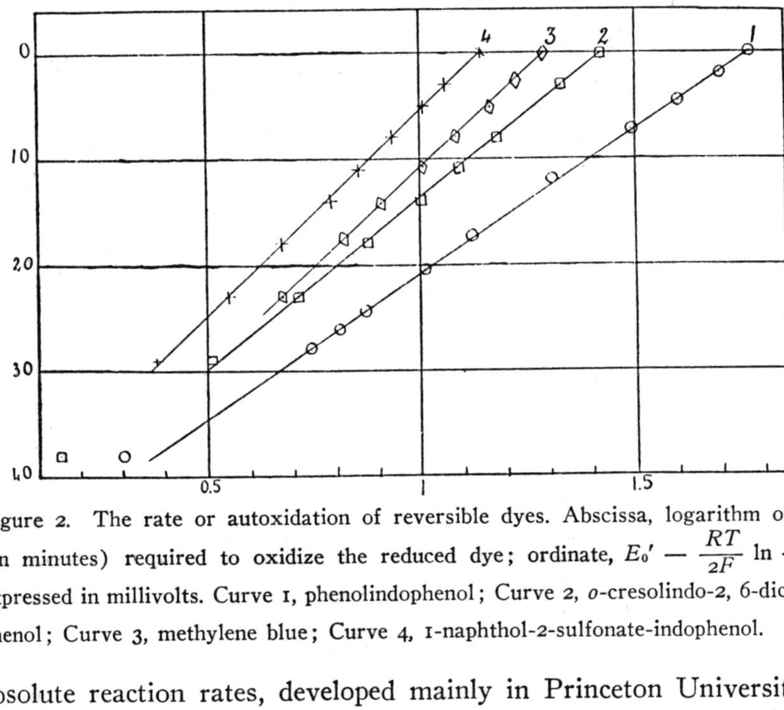

Figure 2. The rate or autoxidation of reversible dyes. Abscissa, logarithm of time (in minutes) required to oxidize the reduced dye; ordinate, $E_0' - \frac{RT}{2F} \ln \frac{a}{1-a}$ expressed in millivolts. Curve 1, phenolindophenol; Curve 2, *o*-cresolindo-2, 6-dichlorophenol; Curve 3, methylene blue; Curve 4, 1-naphthol-2-sulfonate-indophenol.

absolute reaction rates, developed mainly in Princeton University by Eyring and his coworkers (29, 31, 89), which focuses its attention on the thermodynamic probability of molecules entering the transition state, has shown that the speed of reaction depends not on the heat of activation alone but also on the entropy of activation:

$$k = \frac{KT}{h} e^{-\Delta F^*/RT}$$

where ΔF^* is the free energy of formation of the activated state, and k and h the Boltzman and Plank constants, respectively. This equation shows that the velocity constant (k_1), is a function of the work ($-\Delta F^*$) necessary to get the reactants into the transition state. If the concept of the transition state is combined with quantum-mechanical concepts of stabilization by resonance and the dependence of resonance on the number of possible unperturbed forms, we arrive at a rather satisfactory method of dealing with rate problems.

The activated substrate is oxidized by one or more of oxidation-reduction systems, some sluggish, some electromotively active, which transfer electrons to molecular oxygen. If more than one system is interposed between substrate and oxygen, the electron transfer occurs step by step from the system of more negative potential up to molecular oxygen.

A study of the oxidation-reduction systems which take part in the transfer of electrons from substrate to oxygen, shows that these systems where the oxidation involves a total two-electron transfer (pyridine nucleotides, flavins) do so in steps of one electron through radical formation as an intermediate step. The climbing over a large energy hill involved in a two-electron oxidation is made easy by the interposition of the intermediate step. Oxidation of foodstuffs is thus performed by one-electron steps.

Little is still known about the oxidation-reduction potentials of pyridine nucleotides and flavins when in combination with the proteins. Both combine reversibly with proteins, and the dissociation constants of the oxidized and reduced states are sometimes different. However, a model can be provided by a study of the oxidation-reduction potentials of Fe-protoporphyrin and of Fe-protoporphyrin combined reversibly to nitrogenous bases to form the hemochromogens. On addition of co-ordinating bases to the Fe-protoporphyrin there is formation of a series of oxidation-reduction systems which at *p*H 9.20 go from −0.292 volt in Fe-protoporphyrin to +0.050 volt in nicotine hemochromogen (Fig. 3) (5). A similar change of potentials has been observed when flavin is combined to a protein as in Warburg's old yellow ferment, where the potential jumped from −0.208 volt at *p*H 7.0 for flavin to −0.061 volt for the latter (47). W. M. Clark and his coworkers (22) have given a quantitative treatment of the equilibrium reactions between a nitrogenous base and Fe-porphyrin and have shown the relation between the change of potential and the dissociation constants of the metallo-porphyrin-nitrogenous base complex compounds. If association of the Fe^{+++} and Fe^{++} state with the nitrogenous base is equal in strength, there will be no change in potential. If the potential becomes more positive it

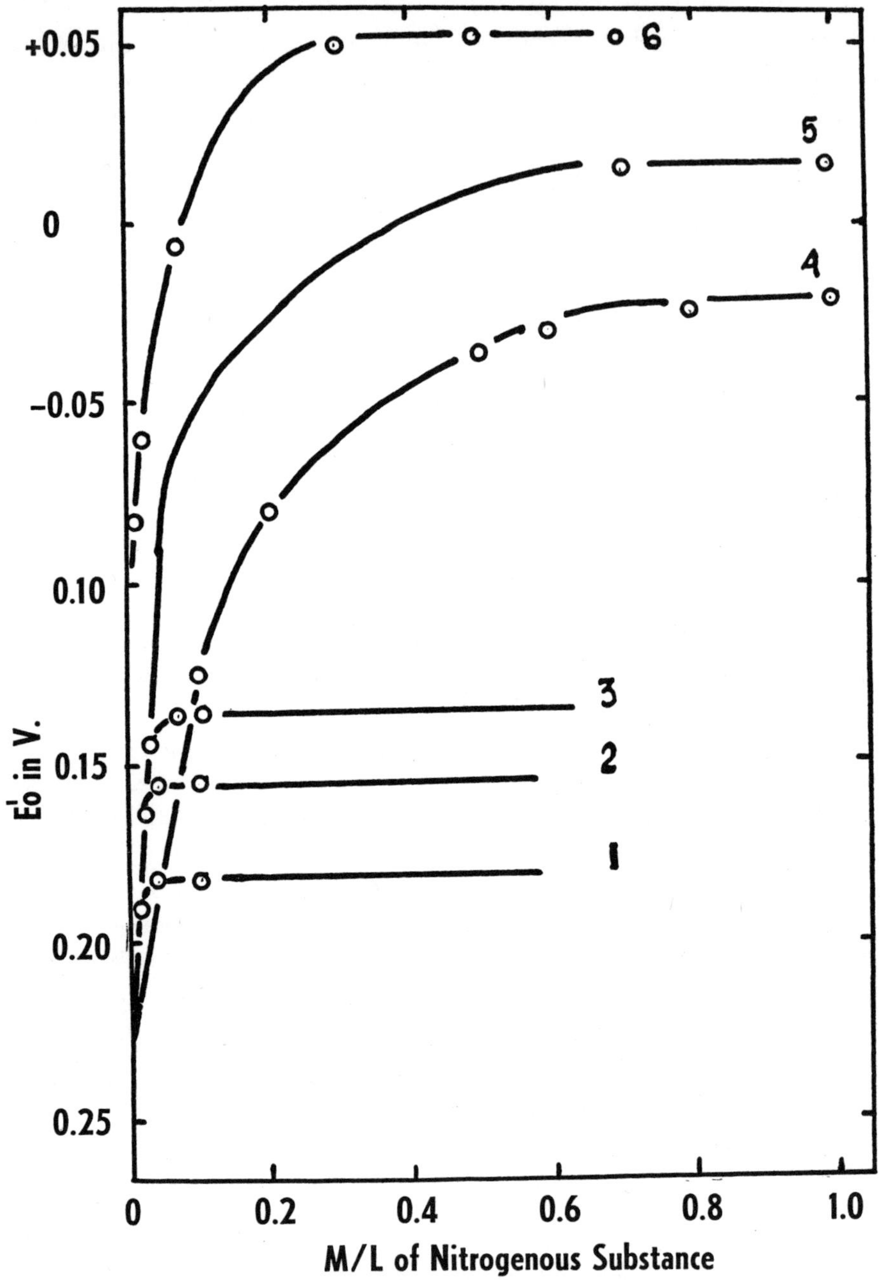

Figure 3. The oxidation-reduction potentials of hemin and of some of its hemochromogens. The influences of the concentration of the following nitrogenous substances: (1) cyanide-hemochromogen; (2) pilocarpine-hemochromogen; (3) histidine hemochromogen; (4) α-picoline-hemochromogen; (5) pyridine-hemochromogen; (6) nicotine-hemochromogen. From Barron, *J. Biol. Chem.*, *121*, 285, 1937.

means that the Fe^{++} porphyrin forms the more stable compound; if the Fe^{+++} form associates more strongly than the Fe^{++} form the trend of the potential will be in the negative direction. This is of great significance for an understanding of the varied catalytic effects of the different oxidation-reduction systems which form part of the enzyme systems. In the same manner as the Fe-porphyrins, and its nitrogenous complexes give a series of oxidation-reduction systems, the potentials of the pyridine nucleotides and flavins in combination with different proteins may change, at constant *p*H, according to the value of the dissociation constants of the protein complex.

The breakdown of foodstuffs in stepwise manner is facilitated by the interposition of reversible oxidation-reduction enzymatic systems, and the stepwise transfer of electrons is facilitated by the property of the three main electron-transfer systems (pyridine nucleotides, flavins, Fe-porphyrins) of forming reversible complex compounds with proteins with a wide variety of oxidation-reduction potentials. It must be recalled, however, that electron transfer between two reversible enzymatic systems does not occur but through the mediation of electroactive systems as shown by the experiments of Borsook (Table IX) (17).

The Pyruvate Oxidative Pathway

Of all the foodstuffs, carbohydrate provides most free useful energy in its breakdown. It is mainly carbohydrate-provided energy which is used during the process of embryonic growth and development. In general, carbohydrate breakdown occurs in two phases: the anaerobic phase which ends in pyruvic acid formation, and the oxidative phase which involves the utilization of pyruvate *via* the oxidative pathway. In 1932 Barron and Miller (11) found that pyruvate was oxidized to acetate and CO_2 by gonococci. Although formation of acetate from pyruvate by animal tissues has also been reported (46, 84, 55), there is a tendency to ignore this oxidative pathway of pyruvate metabolism and to give as the main oxidative pathway that of pyruvate carboxylation to oxaloacetate and condensation between pyruvate and oxaloacetate, as indicated by Krebs in his "tricarboxylic acid cycle" (45). The powerful inhibiting effect of fluoroacetate on the oxidation of acetate was used by Bartlett, Ahrens, and Barron (14) to test the oxidative pathway of pyruvate metabolism. To discover this oxidative pathway fluoroacetate is indeed the ideal agent, since if acetate is formed during the metabolism of pyruvate, and it normally disappears in condensation reactions (acetyla-

TABLE IX

Example of Reversed Reaction.[1]

$\text{Formate}^- + H_2O + \text{Pyruvate} \rightarrow HCO_3^- + \text{Pyruvate}$

pH = 7,0 (phosphate buffer); temp. = 37°, time = 24 hrs., total volume = 10 cc.

Tube no.	Formate micro mols.	Pyruvate micro mols.	Toluene treated organisms	Boiled organisms	Dye 1 micro mol.	E_0' of dye	Lactate formed milliequivalent
1	20	20	+	—	Phenol-indophenol	+0.23	0.2
2	20	20	+	—	Methylene blue	+0.010	0.1
3	20	20	+	—	Indigo monosulphonate	—0.16	0.5
4	20	20	+	—	Cresyl violet	—0.17	1.0
5	20	20	+	—	Janus green	—0.26	10.0
6	20	20	+	—	Methylene violet	—0.26	12.0
7	20	20	+	—	Rosindulin 2 G	—0.28	1.5
8	20	20	+	—	Neutral red	—0.34	2.8
9	20	20	+	—	—	—	0
10	20	20	—	—	Methylene violet	—0.26	0
11	20	20	—	—	—	—	0
12	—	20	+	—	Methylene violet	—0.26	0
13	20	—	+	—	Methylene violet	—0.26	0
14	—	—	+	—	—	—	0
15	20	20	+	+	Methylene violet	—0.26	12.0
16	20	20	—	+	Methylene violet	—0.26	0.6
17	20	20	+	+	—	—	0

$\text{Lactate}^- \rightarrow \text{Pyruvate}^- + 2H^+ + 2(e);\ E_0' = -0{,}180$

$\text{Formate}^- + H_2O \rightarrow HCO_3 + 2H^+ + 2(e);\ E_0' = 0{,}41$

[1] From Borsook (17).

tions, citrate formation, fatty acid synthesis), it will not be detected unless some of these reactions are inhibited. Experiments where the metabolism of pyruvate was studied in the presence of fluoroacetate in kidney, liver, and muscle showed accumulation of acetate and inhibition of pyruvate utilization due to accumulation of its oxidation product, acetate. The other pathways of pyruvate metabolism (amination to alanine, reduction to lactate, and dismutation to lactate and acetate) were not affected. Furthermore none of the oxidative steps of the tricarboxylic acid cycle—oxidation of isocitric, α-ketoglutaric, succinic, and malic acid—were affected by fluoroacetate. The experiments with fatty acids showed an inhibition in the oxidation and utilization of these substances with an inhibition of acetoacetate formation in liver slices, but with acetate as substrate there was accumulation of acetoacetate. This inhibition of acetoacetate formation in the presence of fluoroacetate supports the view, now generally accepted, that the oxidation of fatty acids occurs by steps until the formation of acetate. In the metabolism of acetate there are then two pathways: oxidation inhibited by fluoroacetate, and condensation to acetoacetate not affected by it. Since acetate accumulates in the presence of fluoroacetate, it is plausible to postulate that acetate is the end product in the oxidation of fats and carbohydrates. Furthermore the high toxicity of fluoroacetate is indication of the essential role of acetate metabolism in higher animals and in some plants.

The oxidative pathway, however, is not the only avenue of utilization of pyruvate. Pyruvate, the most highly reactive molecule among the intermediates found during the breakdown of foodstuffs, is in fact utilized in a variety of ways in the different cells, and even in the same cell the utilization of pyruvic acid will depend on the oxygen tension, thiamin concentration, glutathione concentration, etc. The multiple pathway of pyruvate metabolism postulated in this laboratory for several years may thus be summarized as follows, all these reactions being reversible (see chart).

How much of the oxidative removal is performed *via* acetate and how much *via* carboxylation to oxaloacetate has not been determined in all tissues. In the experiments of Bartlett and Barron (14) with kidney slices, at least 45 per cent of the pyruvate utilized could be accounted for as due to direct oxidation.

It can be seen from this chart that the only difference between this scheme and Krebs' tricarboxylic acid cycle is that here pyruvate is directly oxidized to acetate. The synthesis of citric acid from acetate was demonstrated in yeast by Sonderhoff and Thomas (79). The oxidation

The Multiple Pathway of Pyruvate Metabolism

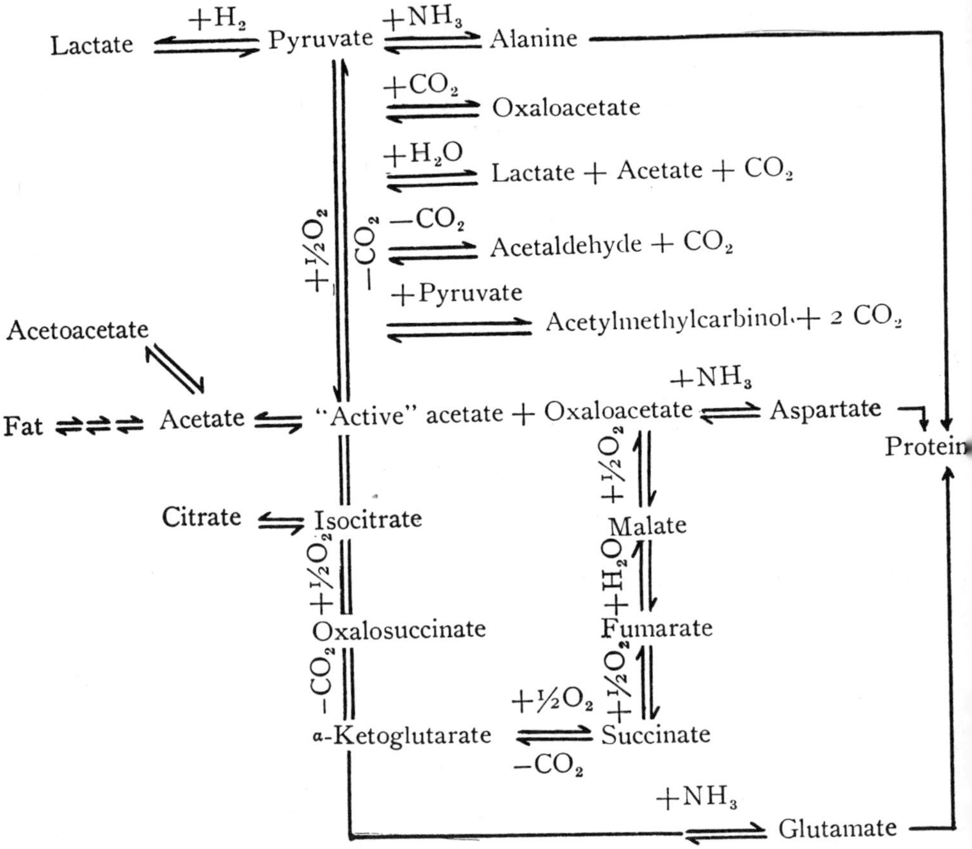

of isocitric acid by isocitric dehydrogenase and triphosphopyridine nucleotide (TPN) to oxalosuccinic acid and its decarboxylation to α-ketoglutaric acid by a specific carboxylase has been demonstrated by the beautiful experiments of Ochoa (67, 69). The other systems have been known for some time.

The interposition of enzymatic reversible oxidation-reduction systems in the various steps during the burning of foodstuffs, already shown in the scheme of pyruvate metabolism, allows the production of a series of coupled reactions whose speed and direction will be determined not only by the value of the equilibrium constants but also by the environment (temperature, O_2 tension, CO_2 tension, electrolytes, hormones, etc.).

Ochoa (68, 69) has thus shown a number of coupled reactions which in *in vitro* experiments end in carboxylations:

(1) α-ketoglutarate + CO_2 + glucose 6-phosphate $\rightleftharpoons$ 1-isocitrate + phosphogluconate;

(2) Malate + TPN $\rightleftharpoons$ pyruvate + CO_2 + $TPNH_2$

These reactions do not occur in the presence of pure isolated enzyme systems. Carboxylation of α-ketoglutarate and pyruvate, difficult of attainment because of the values of the equilibrium constants, was rendered possible by coupling carboxylation with an oxidation-reduction reaction. Whether such reactions do occur in living cells is problematic.

Oxidations and Phosphorylations

Whenever an important discovery is made, there spring forth applications, similarities, extensions. Years have to pass by before harvesting from the melee the useful crop. The coupling between phosphorylation and oxidation observed by Meyerhof (59) in his studies on the mechanism of glycolysis was explained in 1939 by the striking demonstrations in Warburg's laboratory (83) which culminated in 1942 in the isolation and crystallization of the specific transphosphorylase (18). Warburg and his coworkers showed that the aldehyde group in phosphoglyceraldehyde is not oxidized as aldehyde hydrate but as aldehyde phosphate. By a non-enzymatic reaction glyceraldehyde 3-phosphate takes up another phosphate group to give 1,3-diphosphoglyceraldehyde, which is oxidized by DPN to 1,3-diphosphoglyceric acid:

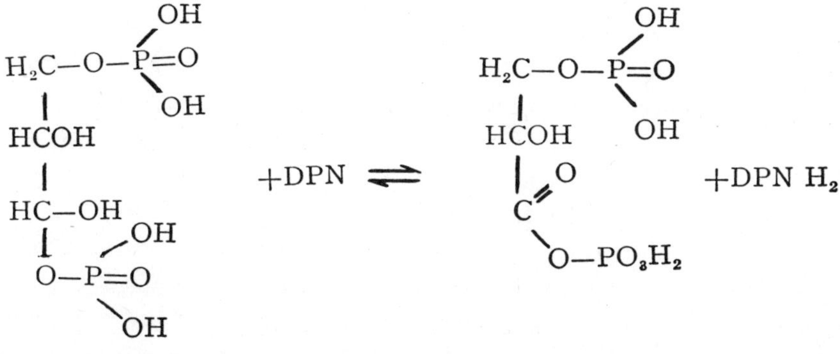

The phosphate transfer enzyme, a Mg-protein, transfers the unstable phosphate group into adenosinediphosphate to give ATP. This reaction has been taken as a model of phosphate transfer from aceylphosphates:

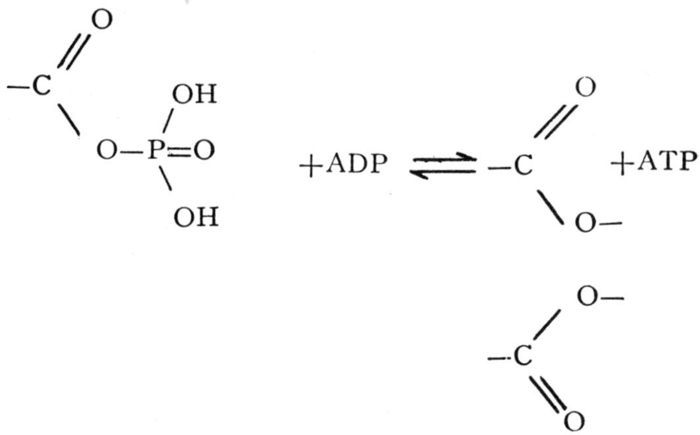

The resonating structures impart stability to the group (39).

Lipmann (53) has applied and extended these concepts to the oxidation of pyruvate and α-keto acids in general, and indeed has found that certain bacteria produce acetylphosphate on oxidation of pyruvic acid, and that animal tissues hydrolyze acetylphosphate rather rapidly (54). These important contributions of Lipmann were promptly accepted, but little has been accomplished towards further clarification of the mechanism of aerobic phosphorylations. The fact that they exist was demonstrated by Belitzer and Tsibakowa (15), Ochoa (65) and Colowick *et al.* (24). From Ochoa's work it is clear that on the oxidation of pyruvate there are about 15 phosphorylations per mole of pyruvate completely oxidized, i.e. 3 per atom of oxygen. Of these 15 phosphorylations, Ochoa (66) has reported that during the oxidation of α-ketoglutaric acid to succinic acid *in the presence of glucose and adenylic acid,* 3 phosphate molecules are simultaneously esterified to the sugar forming hexose diphosphate. The transfer of 2 electrons to oxygen was thus accompanied by the release of energy in three successive steps in the form of high-energy phosphate bonds. How they are formed remains unknown. The postulated succinylphosphate formed as the first product of oxidation has not been demonstrated. Ball's suggestion (2) that the step by step transfer of electrons from oxidizable substrate to pyridine nucleotide, from this to flavin, from flavoprotein to cytochrome, and from cytochrome to oxygen might release the energy (0.25 V.) necessary to form a high-energy phosphate bond seems reasonable, but again is still in the realm of speculation. To complicate the problem, Stumpf *et al.* (81) maintain that α-ketoglutarate is oxidized to succinate by preparations from pigeon breast muscle without the necessity of inorganic phosphate, adenosinetriphosphate, and magnesium, all components considered by Ochoa essential for the oxidation process. The phosphorylations in the first step of the cycle (see chart)—pyruvate + $\frac{1}{2}O_2 \rightarrow$ acetate + CO_2—have not been established in animal tissues in spite of the repeated efforts of Lipmann (54). Here again, Stumpf *et al.* maintain that this oxidation occurs in the absence of phosphate and with no formation of acetyl phosphate. The second oxidation step, isocitrate + TPN $\rightleftharpoons$ oxalosuccinate + TPN H_2, is postulated to give no high-energy phosphate bonds; however in the experiments of Ochoa (69) there was but one step in the transfer of electrons. If Ball's suggestion is verified, here as well as in the oxidation of pyruvate, the step by step release of energy might be utilized for the formation of high-energy phosphate bonds. In the fourth step, succinate + $\frac{1}{2}O_2 \rightleftharpoons$ fumarate + H_2O, there has been found evidence of one phosphorylation (66). This is probably due to the high potential of the system (0 V. at *p*H 7) and the fewer steps in electron transport to molecular oxygen (0.8 V.). The final step, malic + $\frac{1}{2}O_2 \rightleftharpoons$ oxaloacetic + H_2O, was found by Kalckar (40, 41) to produce at least two phosphorylations. It seems as if the insistence on finding analogs to Warburg's reaction, which after all is a fermentation reaction, has stopped progress and hindered the search for other channels of energy transfer. Perhaps Ball's suggestion ought to be studied experimentally. Aerobic phosphorylations may be produced by the release of energy from the different oxidation-reduction catalysts to adenylic acid, for it is accepted that adenylic acid enhances completeness of oxidation. Lack of phosphorylation in isolated enzyme systems where substrate is oxidized by only the first oxidation-reduction system and failure of phosphorylation in some tissue extracts when cytochrome C is not present favor this opinion. The ATP thus produced would represent the free useful energy ready to be used.

Energy transfer from electrons and its utilization by the cell through adenosinetriphosphate occurs not only on the oxidation of carbohydrates but also on that of fats, as the studies of Lehninger (50) and Muñoz and Leloir (64) have demonstrated.

The Oxidation and the Synthesis of Fats

Our knowledge of the mechanism of oxidation and synthesis of fatty acids has advanced considerably in the last years through the efforts of a number of investigators, although no single enzyme has yet been isolated. The classical concept of Knoop (43) and Dakin (25) with the modifications suggested by MacKay (57)—oxidation at their βC atoms with subsequent removal of two C units which may condense to acetoacetate or to citrate—seems now generally accepted. As in the complete oxidation of carbohydrate, the series of oxidations which take place during the breakdown of fatty acids require the presence of ATP or of C_4 dicarboxylic acids, which in their oxidation provide energy-rich phosphate bonds (50, 64, 38). Whether there is a transphosphorylation process with the formation of unstable acyl phosphates is not known, in spite of some efforts to prove that such substances are formed. The acyl phosphates (monopalmityl phosphoric acid and monooctanoylphosphoric acid), although hydrolyzed much more slowly than acetyl phosphate (51), were physiologically inactive. When octanoic acid was oxidized by the liver in the absence of C_4 dicarboxylic acids, acetoacetate was the end product (52, 85). In the presence of C_4 dicarboxylic acid, the C_2 fragments gave citric acid and its oxidation products, α-ketoglutaric acid, and succinic acid. The interconversion of acetic acid to acetoacetic acid has been demonstrated by direct interconversion (49, 86) as well as by the synthesis of citric acid from acetoacetate presumably by its previous breakdown to "active" acetate (87). This acetic acid, the end product of carbohydrate oxidation and of fat oxidation, is the connecting link of carbohydrate and fat metabolism. The further oxidation of acetic acid, we have already shown (see chart), occurs through the tricarboxylic acid cycle. The synthesis of fat starts with the condensation of acetate to acetoacetate. Indeed, Rittenberg and Bloch (75) demonstrated with isotope experiments that acetic acid is utilized for the synthesis of fats.

The Oxidation and Synthesis of Proteins

Proteins in the oxidation of their fundamental units, the amino acids, contribute greatly to the formation of the carboxylic acids which by their reversible coupled oxidation-reductions form the common pathway for the final steps of complete oxidation of foodstuffs, whether it be carbohydrate, fat or protein. The rapid and continuous breakdown and synthesis of proteins so well demonstrated by Whipple and his coworkers (56) and by Schoenheimer and his coworkers (77), as well as the great specificity of proteinases and peptidases, brought forth the suggestion that protein synthesis could be performed by coupled reactions between the hydrolytic enzymes. Bergmann and Fruton (16), in fact, presented models of peptide bond synthesis by selecting suitable amino acid derivatives and by producing the conditions necessary for reversal of hydrolysis. Whether this mechanism actually takes place in living cells is not known. The synthesis of proteins, like that of glycogen, must be due to the cooperation of a large number of enzymes. In cytochrome containing living cells it does not occur in the absence of oxygen, and it is plausible to postulate that proteins are, like carbohydrates and fats, synthesized after their breakdown to amino acids and that peptide bond formation requires the utilization of energy produced during the oxidation of carbohydrates and fats. Support for this assumption is given by the experiments of Cohen and McGilvery (23) on the formation of *p*-amino-hyppuric acid from *p*-aminobenzoic acid and glycine by rat tissue:

$$NH_2C_6H_4COOH + NH_2CH_2COOH \rightarrow NH_2C_6H_4CO{-}NHCH_2COOH + H_2O$$

This may be taken as a model of peptide bond synthesis. This reaction could not proceed anaerobically nor in the presence of oxidative inhibitors (HCN, NaN_3). Inhibition of this reaction by arsenite, iodoacetate and malonate favors the view that the energy for the synthesis is provided by oxidations in the Krebs cycle where there are a number of —SH enzymes. Whether this energy is furnished by the high-energy phosphate bonds produced in these oxidations is not yet established. It must be emphasized however that proteins possess a great specificity and complexity. Besides enzymes forming peptide bonds there must be others which add these bonds together to form the complicated and varied structures of the different proteins.

Casperson (19, 20, 21) demonstrated that ribonucleic acid is present in abundance in growing cells and gave some evidence in favor of the view that it is essential for protein synthesis. The mechanism of action of nucleic acids is not yet known. They must act in the last steps when the architecture of the molecule acquires its specific characteristics. Spiegelman and Kamen (80) believe that they are "the specific energy donators which make possible reactions leading to protein and enzyme synthesis." They found that new protein formation in yeast cells was accompanied by a marked transfer of phosphate from the nucleoprotein. However the phosphate groups in nucleoproteins are phosphate esters which on hydrolysis provide small amount of energy. Northrop (64a) has dealt extensively on this subject of the last stages of protein synthesis.

Regulatory Mechanisms

From this brief review of some enzyme reactions it can be seen that what distinguishes biochemical reactions from chemical reactions in general is the means provided for the rapid attainment of equilibrium, so strikingly demonstrated by the enzymatic reversible oxidation-reduction systems. Furthermore, although on a thermodynamic basis a very large number of reactions are theoretically possible in a given biologic system, the presence of enzymes will result in the selective catalysis of only a limited number of these reactions. But the isolation, purification, and study of the catalytic action of individual enzymes is only the beginning of the study of the biochemical reactions which occur in the cell, whether it be in the state of "dynamic equilibrium" or in the state of growth. Living matter is the result of the coordinated interplay of substance and structure, as Pauling (71) has remarked, and the advance of biology requires equal devotion to the work of isolation and identification of active chemical substances, to the study of the macromolecular stromatic structures present in the living cell, and to the investigation of the multiple regulating mechanisms which contribute to the cellular dynamic equilibrium.

The rate of reaction of isolated enzyme systems is extremely high when compared with the rate of reaction in the living cell. Enzyme reactions are controlled in the living cell by a variety of regulating mechanisms. To the usual physico-chemical factors which govern the rate of reactions in solutions there must be added the effect of macromolecular structures which constitute the framework of the cell, such as the genes, and which can orient or fix the position of the intracellular enzymes by means of loose unions. As the cell increases in efficiency and becomes more differentiated, the activity of the enzymes becomes more and more controlled by specific regulating mechanisms, which in multicellular organisms freely circulate through the intercellular fluid. In vertebrates these regulatory mechanisms increase in number and specificity, and they control and orient the rate of enzymatic reactions by interaction among each other. These regulatory mechanisms (hormones, nucleoproteins, genes, electrolytes, certain oxidation-reduction systems) do not act by forming part of enzyme components; they act by controlling the rate of enzyme reactions or by changing the orientation of these reactions. They act as the traffic policemen in the crowded streets of a city. Thus the steady state of the adult cell is maintained, and the synthetic processes required in growing cells are brought forth. An example of the presence of regulating mechanisms is seen in the effect of dilution on the respiration of sea urchin sperm. The respiration of sea urchin sperm is maintained at a low level under the regulating influence of some substances possessing —SH groups; on dilution or on addition of iodoacetate or malonate (usual respiratory inhibitors), the regulatory system becomes inactivated and the respiration is increased (Fig. 4) (9). A striking demonstration of the role of hormones as regulators of enzyme reactions, a theory postulated in 1943 (6), is found in the experiments of Price, Colowick, and Cori (72) of inhibition of muscle hexokinase by a substance (hormone?) contained in the anterior pituitary gland and the release of this inhibition by insulin. In the living cell enzymes may be present which are so located as to be inactive in the intact cell. Physiological alterations, rupture of the cell, destruction of the normal architecture, may bring these enzymes into action. An example of physiological alteration is given by the experiments of Korr (44) who found that the respiration of the salivary glands, which is insensitive to azide when the gland is in the resting state, becomes azide sensitive when the gland is stimulated. An example of enzyme liberation on damage to the cells is given by the marked and rapid inhibition of the respiration of lung tissues after the tissue has been ground (7). The QO_2 value goes from 7.5 down to 1.0. This inhibition is due to destruction of the pyridine nucleotides by the nucleosidase present in lung tissue in high concentration but kept inactive in the intact cell. Evans (28) has dealt with the properties of the anterior hypophyseal growth hormone isolated in his laboratory which plays an important role in the synthesis of proteins, a role which in our opinion is that of a regulating agent.

Cell Division

From the extensive investigations on the mechanism of cell division it must be concluded that there is as yet no full understanding of the physical and chemical mechanisms which bring forth this process. There is agreement however that we are confronted with a series of events which together culminate in the phenomena of membrane formation and cell division. The initial process, formation of fertilization membrane and entrance of the spermatozoon into the ovum with no nuclear changes, may be called the *anaerobic phase*, because it may take place under strictly anaerobic conditions (3). If energy is required for this process it must be obtained from glycolysis. The nuclear changes which

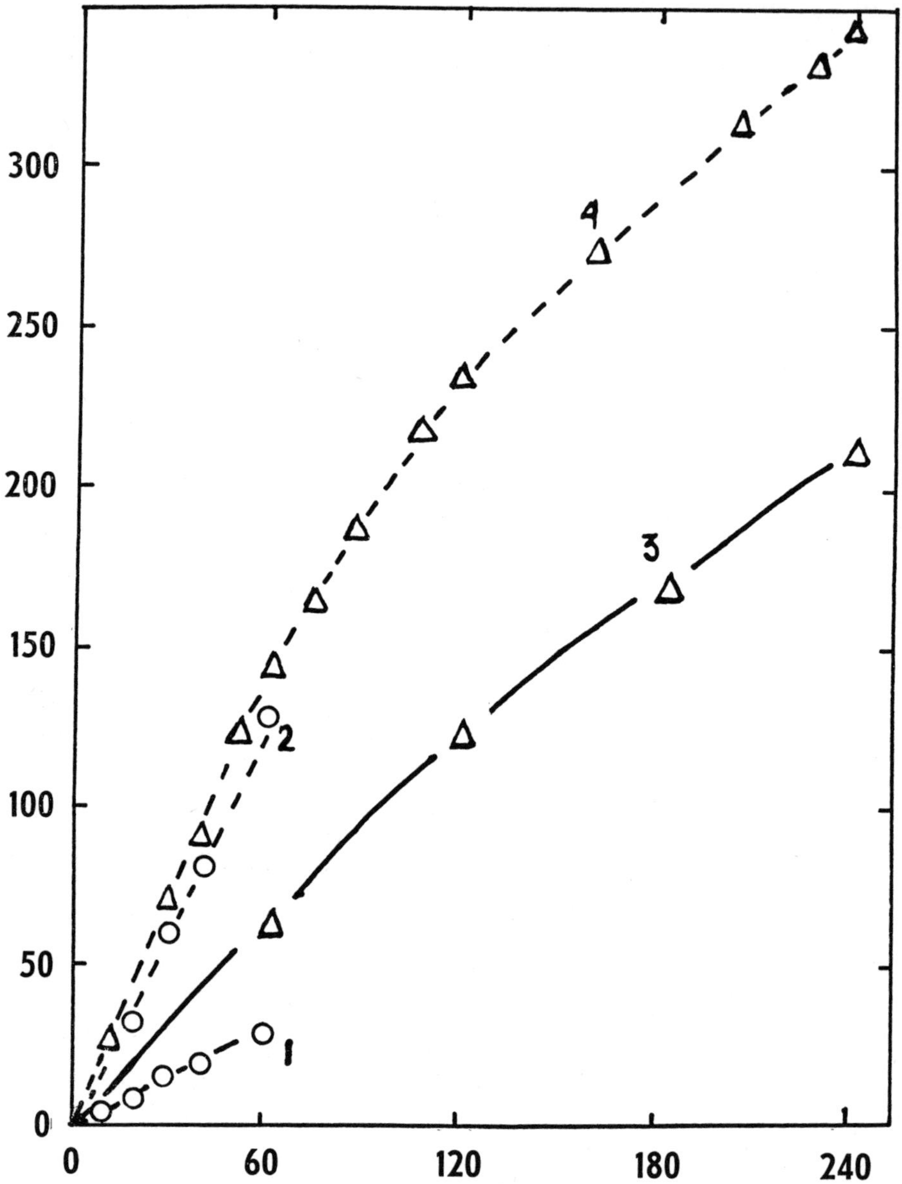

Figure 4. Effect of iodoacetate on the O_2 uptake of sea urchin sperm. Abscissa, time in minutes. Ordinate, O_2 uptake in mm^3. (1) Control (heavy suspension). (2) Heavy suspension with iodoacetate (0.001M). (3) Control (dilute suspension). (4) Dilute suspension with iodoacetate (0.001M).

follow membrane formation and end in cell division constitute the *aerobic phase* and must require the energy provided by oxidative reactions. The increased utilization of glycogen accompanied by the increased utilization of pyruvate (Table X) may provide the energy necessary for the process of cell division (32). Furthermore, the introduction during the cell division process of three of four oxidative steps (the series of oxidations carried on by the cytochrome system) in the series of reactions from substrate to molecular oxygen will increase the efficiency of the reactions so that for the same amount of O_2 utilized there will be a more efficient utilization of energy. Whether carbohydrate is metabolized entirely *via* glycolysis, or is oxidized through the pathway of hexosemonophosphate oxidation as suggested by Örström and Lindberg (70) is not known. Neither is the nature of the acid formed on cell division known. Some of it may be acetic acid formed on oxidation of pyruvic acid. Whether acetate is utilized or not is not known. If it is, the pathway of its oxidation must be different from that of most animal tissues because neither succinate nor α-keto-

TABLE X

Utilization of Pyruvate by Arbacia Eggs—Influence of Fertilization.

Date (1941)	Fertilization (%)	Duration (hrs.)	Pyruvate utilized per gm. dry weight per hr. (micrograms)
July 17	80	1	400
July 19	75	2	500
July 21	80	5	345
July 23	88	3	480
July 24	80	0.5	500
July 21	none	1	60
July 23	none	5	70
July 24	none	2	80
July 17	none	1	50
July 19	none	3	69

glutarate is oxidized by sea urchin eggs. Fluoroacetate had no effect on the O_2 uptake.

Cell Growth

Our knowledge of the enzymatic reactions which take part in cellular growth is meager. In general cellular growth, new protein formation, takes place at the expense of aerobic oxidations except in those cells deprived of cytochromes.

Voegtlin (82), Hammett (33), and Rapkine (73) have for a long time championed the importance of sulfhydryl groups in cell growth, and evidence in support of this opinion is good. They have been found increased in concentration in growing cells (62, 88) and growth was inhibited on addition of substances which combine with —SH groups (27, 34, 74, 76). It has already been pointed out that there are a large number of enzymes which require the presence of —SH groups in their protein moiety for activation. Destruction of these groups by oxidation (x-rays, O_2, Cu), alkylations (CH_2ICOOH, $CH_2ICOONH_2$, mustard, nitrogen mustard), or mercaptide formation (Hg, Pb, Cd, Zn, Bi, As, V, Sb) will produce enzyme inhibitions, an inhibition which affects such high-energy yielding reactions as pyruvate oxidase, α-ketoglutarate oxidase, succinoxidase, triosephosphate oxidase, adenosinetriphosphatase, malate oxidase. Sluggish oxidation-reduction systems, mostly in their reduced state (glutathione, ascorbic acid), have also been found abundant in cells during growth and development. These systems must have for function that of keeping the —SH groups of the enzymes in the reduced, physiologically active state.

It must be recalled that ionizing radiations (63), halogenated alkylamines (30), and mustard gas (1) inhibit mitosis. Furthermore Hevesy (36) has found that the synthesis of nucleoproteins (which take part in mitosis) is inhibited by x-rays. Under certain conditions enzymes like adenosinetriphosphatase from aged myosin are so susceptible to ionizing radiations that amounts as small as 1r produce inhibition (Table XI). The inhibiting effect of x-rays may well indicate that some enzyme necessary for the synthesis of nucleoproteins requires —SH groups for its activity, groups which are oxidized by the OH and O_2H radicals, and H_2O_2 produced during the irradiation of water. The inhibiting effect of mustard gas and of nitrogen mustard might also be due to combination of these substances with the —SH groups. The fact that lewisite produces no effect (1) is no serious objection, because

Barron *et al.* (12) have shown that *d*-amino acid oxidase, a sulfhydryl enzyme, was not inhibited by lewisite.

The close interrelation between the metabolism of carbohydrates, fats, and proteins has shown that for the synthesis of protoplasm energy is obtained by the aerobic oxidation of carbohydrate. The preponderant

TABLE XI

The Effect of Very Low Doses of X-Ray on Myosin (Aged).

	Phosphorus liberated in 10 min. at 38° (micrograms)		
Dosage (r)	Treated	Control	Inhibition (%)
10	5.7	11.9	52
5	5.4	11.9	55
1	8.4	11.9	30

part of synthesis reactions over breakdown reactions in growing cells must be due not only to the increased formation of enzymes but also to the distribution of the regulating mechanisms which not only modify the rate of reactions but are also able to change the direction of these reactions. The study of these regulating mechanisms which channel the multiple enzymatic reactions towards different pathways will clarify the role of enzymes in growth processes and will help us in understanding the change from normal controlled growth to the uncontrolled growth of cancer.

REFERENCES

1. Auerbach, C., J. M. Robson, and J. G. Carr. *Science, 105*, 247, 1947.
2. Ball, E. G. *Ann. New York Acad. Sci., 45*, 363, 1944.
3. Barron, E. S. G. *Biol. Bull., 62*, 42, 1932.
4. Barron, E. S. G. *J. Biol. Chem., 97*, 287, 1932.
5. Barron, E. S. G. *J. Biol. Chem., 121*, 285, 1937.
6. Barron, E. S. G. *Adv. Enzymology, 3*, 149, 1943.
7. Barron, E. S. G., Z. B. Miller and G. R. Bartlett. *J. Biol. Chem., 171*, 791, 1947.
8. Barron, E. S. G., S. R. Dickman, and T. P. Singer. *Feder. Proc. Soc. Biol. Chem., 6*, 1, 1947.
9. Barron, E. S. G., and J. M. Goldinger. *Proc. Soc. Exp. Biol. and Med., 48*, 570, 1941.

10. Barron, E. S. G., and L. A. Hoffman. *J. Gen. Physiol., 13*, 483, 1930.
11. Barron, E. S. G., and C. P. Miller. *J. Biol. Chem., 97*, 691, 1932.
12. Barron, E. S. G., Z. B. Miller, G. R. Bartlett, J. Meyer, and T. P. Singer. *Biochem. J., 41*, 69, 1947.
13. Barron, E. S. G., and T. P. Singer. *J. Biol. Chem., 157*, 221, 1945.
14. Bartlett, G. R., R. Ahrens, and E. S. G. Barron. *J. Biol. Chem., 170*, 67, 1947.
15. Belitzer, V. A., and E. T. Tsibakowa. *Biokhimiya, 4*, 516, 1939.
16. Bergman, M., and J. S. Fruton. *Advances in Enzymology, 1*, 63, 1941; *Ann. New York Acad. Sci., 45*, 409, 1944.
17. Borsook, H. *Ergilo. Enzymforsch., 4*, 1, 1935.
18. Bucher, H. *Naturwiss., 30*, 756, 1942.
19. Casperson, T. *Chromosoma, 1*, 562, 1940; *Naturwiss., 17*, 33, 1941.
20. Casperson, T., and K. Brandt. *Protoplasma, 35*, 507, 1940.
21. Casperson, T., and L. Santerson. *Acta Radiol. Supp., 46*, 1942.
22. Clark, W. M., J. F. Taylor, T. H. Davies, and C. S. Vestling. *J. Biol. Chem., 135*, 543, 1940.
23. Cohen, P. P., and R. W. McGilvery. *J. Biol. Chem., 166*, 261, 1946.
24. Colowick, S. P., M. S. Welch, and C. F. Cori. *J. Biol. Chem., 133*, 359, 641, 1940.
25. Dakin, H. E. *J. Biol. Chem., 6*, 221, 1909.
26. Dale, W. M. *Biochem. J., 34*, 1367, 1940.
27. Ephrusi, B. *Compt. Rend., 192*, 1763, 1931.
28. Evans, H. M. *Am. Scientist*, Vol. 35, p. 465, 1947.
29. Eyring, J. *J. Chem. Physics, 3*, 107, 1935.
30. Friedenwald, J. Personal communication, 1943.
31. Glasstone, Laidler, and J. Eyring. *The Theory of Rate Processes.* McGraw-Hill Book Co., 1941.
32. Goldinger, J. M., and E. S. G. Barron. *J. Gen. Physiol., 30*, 73, 1946.
33. Hammett, F. S. *Cold Spring Harbor Symp. Quant. Biol., 2*, 78, 1934.
34. Hammett, F. S. *Protopl., 7*, 297, 1929.
35. Hellerman, L., F. P. Chinard, and V. R. Dietz. *J. Biol. Chem., 147*, 443, 1943.
36. Hevesy, G. *Rev. Med. Physics, 17*, 162, 1945.
37. Hopkins, F. G., and E. J. Morgan. *Biochem. J., 32*, 611, 1938.
38. Hunter, F. E., and F. L. Leloir. *J. Biol. Chem., 159*, 295, 1945.
39. Kalckar, H. *Ann. New York Acad. Sci., 45*, 395, 1944.
40. Kalckar, H. *Ann. Rev. of Biochem., 14*, 283, 1945.
41. Kalckar, H. *Enzymologia, 2*, 47, 1935; *5*, 365, 1939.
42. Kalnitsky, G., and E. S. G. Barron. *J. Biol. Chem., 170*, 23, 1947.
43. Knoop, F. *Beit. Chem. Physiol. Path., 6*, 150, 1904.
44. Korr, J. M. *Am. J. Physiol., 133*, 355, 1941.
45. Krebs, H. A. *Advances in Enzymology, 3*, 171, 1943.
46. Krebs, H. A., and W. A. Johnson. *Biochem. J., 31*, 645, 1937.
47. Kuhn, R., and P. Boulanger. *Ber. Chem. Ges., 69*, 1557, 1936.
48. Lardy, H. A., and J. A. Ziegler. *J. Biol. Chem., 159*, 343, 1945.
49. Lehninger, A. *J. Biol. Chem., 143*, 147, 1942.

50. Lehninger, A. *J. Biol. Chem., 157,* 363; *161,* 437, 1945.
51. Lehninger, A. *J. Biol. Chem., 162,* 333, 1946.
52. Lehninger, A. *J. Biol. Chem., 164,* 291, 1946.
53. Lipmann, F. *Advances in Enzymology, 1,* 99, 1941.
54. Lipmann, F. *Advances in Enzymology, 6,* 231, 1946.
55. Long, C. *Biochem. J., 32,* 1711, 1938.
56. Madden, S. C., and G. H. Whipple. *Physiol. Rev., 20,* 194, 1940.
57. MacKay, E. M. *J. Clin. Endoc., 3,* 101, 1943.
58. Meyerhof, O., P. Ohlmeyer, W. Gentner, and H. Maier-Leibenitz. *Biochem. Z., 298,* 376, 1938.
59. Meyerhof, O., P. Ohlmeyer, and W. Möhle. *Biochem. Z., 297,* 113, 1938.
60. Michaelis, L. *Transact. Electrochem. Soc., 71,* 107, 1940.
61. Michaelis, L., and M. L. Menten. *Biochem. Z., 49,* 333, 1913.
62. Miller, B. J., and S. P. Reimann. *Arch. Path., 29,* 181, 1940.
63. Müller, H. J. *Verh. Vint. Kong. Verarbungsw., 234,* 1927.
64. Muñoz, J. M., and L. F. Leloir. *J. Biol. Chem., 147,* 355, 1943.
64a. Northrop, J. This symposium.
65. Ochoa, S. *J. Biol. Chem., 138,* 751, 1941; *151,* 493, 1943.
66. Ochoa, S. *J. Biol. Chem., 155,* 87, 1944.
67. Ochoa, S. *J. Biol. Chem., 159,* 243, 1945.
68. Ochoa, S., A. Mehler, and A. Kornberg. *J. Biol. Chem., 167,* 871, 1947.
69. Ochoa, S., and E. Weisz-Tabori. *J. Biol. Chem., 159,* 245, 1945.
70. Orström, A., and O. Lindberg. *Enzymologia, 8,* 367, 1940.
71. Pauling, L. *Chem. and Engin. News, 24,* 1375, 1946.
72. Price, W., S. P. Colowick, and C. F. Cori. *J. Biol. Chem., 160,* 633, 1945.
73. Rapkine, L. *Ann. Physiol., 7,* 382, 1931; *J. Chim. Phys., 33,* 493, 1936; *34,* 416, 1937.
74. Rapkine, L., E. Chatton, and A. Lwoff. *Compt. Rend. Soc. Biol., 106,* 626, 1931.
75. Rittenberg, D., and K. Bloch. *J. Biol. Chem., 160,* 417, 1945.
76. Ruffili, D. *Boll. Soc. ital. Biol. sper., 17,* 75, 1942.
77. Schoenheimer, R. *Dynamic State of Body Constituents.* Harvard. Univ. Press, Cambridge, 1942.
78. Singer, T. P., and E. S. G. Barron. *J. Biol. Chem., 157,* 241, 1945.
79. Saunderhoff, R., and H. Thomas. *Ann. der Chemie, 530,* 195, 1937.
80. Spiegelman, S., and M. D. Kamen. *Science, 104,* 581, 1946.
81. Stumpf, P. K., K. Zarudnaya, and D. E. Green. *J. Biol. Chem., 167,* 817, 1947.
82. Voegtlin, C. *Cold Spring Harbor Symp. Quant. Biol., 2,* 78, 1934.
83. Warburg, O., and W. Christian. *Biochem. Z., 303,* 40, 1939.
84. Weil-Malherbe, H. *Biochem. J., 31,* 2202, 1937.
85. Weinhouse, S., G. Medes, and N. F. Floyd. *J. Biol. Chem., 158,* 411, 1945.
86. Weinhouse, S., G. Medes, and N. F. Floyd. *J. Biol. Chem., 155,* 143, 1944.

87. Weinhouse, S., G. Medes, and N. F. Floyd. *J. Biol. Chem., 166*, 691, 1946.
88. Owen, S. E., H. A. Weiss, and L. H. Prince. *Am. J. Cancer, 35*, 424, 1939.
89. Wynne-Jones, and H. Eyring. *J. Chem. Physics, 3*, 492, 1935.

VII. DIFFERENTIAL GROWTH

BY PAUL WEISS[1]

The appearance of different properties in initially equivalent cells is called "differentiation." Growth being only one cellular attribute among many, "differential growth" is to be subsumed under the larger topic of "differentiation," and our discussion therefore will have to center on the latter. Since the accent of this book lies more on prospects than on retrospects, I intend to concentrate on the gaps in our concepts and knowledge of differentiation and point to possible ways of filling them rather than dwell on past achievements, which, though impressive in themselves, are dwarfed by the task that remains to be accomplished.

Biology has discovered the strength it can derive from a more rigorous analysis of biological phenomena by the tools and standards of the physical sciences, and if the study of development is to share the new light with its more alert sister branches, such as physiology, genetics and immunology, to name only a few, it will have to submit to the same reorientation that has proved so beneficial with the others. This reorientation is in the direction of greater precision, objectivity, and consistency in the description and interpretation of phenomena, preferably in terms of physical and chemical order.

It implies the adoption not only of the tools of the physical sciences, but above all of their disciplined methodology. Analytical embryology is still full of problems which it can only settle within its own right and by its own techniques. But in doing this it must apply standards, criteria, and methods of procedure as rigorous as those of the physical sciences. It must strive to replace the abstract formalism and verbalism of its groping infancy by a mature realism, in which a term stands as symbol for a known or knowable entity or relation, rather than as a soothing device of convenient elasticity to pretend knowledge where there is none. We must not allow its analytical principles to be contaminated by relapses into animistic mythology, as we are doing when we charge the unresolved residue of meticulous determinations of physico-chemical properties to "organizers," "inductors," and similar ill-descript agents. Labels must no longer be allowed to pass for content, nor generalities for explanations. What we need most in entering

[1] University of Chicago. Original work referred to in this article was aided by the Dr. Wallace C. and Clara A. Abbott Memorial Fund of the University of Chicago.

the era ahead is the will to drive the analysis of developmental phenomena far beyond the mark where we used to stop to coin a term, and the determination to become as exact, specific, objective, and realistic in the formulation of our problems and in the description of our observations and results as has become the rule in the physical sciences. This is what we mean by reorientation. It is just as important as is the gathering of more data. With this in mind, we turn to the question which forms the core of my assignment: What is differentiation?

What Is Differentiation?

By the standards just advocated, this question can immediately be recognized as far too indefinite to be answered by anything but another generality. For "differentiation" refers not to a single circumscribed natural phenomenon, but is a summary name for a heterogeneous collection of notions about a variety of events associated with development. Consequently, the various statements and generalizations which have been made in the past concerning "differentiation" and "dedifferentiation" are equally heterogeneous and full of contradictions. Only a few authors have deemed it necessary to be explicit about terms of such general currency. In order not to add to the confusion, I shall begin by developing my own definition. Then whatever I shall have to say about "differentiation" will refer to the phenomenon as here defined, and not to anybody else's "differentiation."

A definition of "differentiation" is partly a matter of objective description, partly of plain logic. We shall deal with the logical portion first, because it has been more persistently ignored.

Characters vs. Processes

One source of confusion is the tradition of confounding the process of differentiation with its criteria. We usually tell differentiation by its more conspicuous products; e.g. melanin granules in a pigment cell, myofibrils in a muscle fiber, secretion granules in a gland cell. Our criteria are predominantly morphological and we judge mostly by appearances. In the examples just mentioned, our judgment happens to be well founded. It becomes dubious when based merely on superficial differences of shape, arrangement, or size, and it misleads us completely when we try to reverse the argument and take the absence of obvious differences for evidence of lack of differentiation. To appraise a cell correctly we must take into account all of its properties, not just the ones that impress us optically, and to appreciate the development of these properties we must bear in mind that we are dealing with material systems in space and time, in which whatever is present at any one stage is the result of antecedent *processes*. Thus the conventional "characters" by which we distinguish differentiation are in reality only incidental manufacturing samples of the cellular production plant. And just as one kind of industrial plant differs from another not only by its different end products but by the whole machinery and processing method turning out those products, so a differentiated cell of one kind differs from that of another kind not only in its final equipment (pigment, fibrils, secretions, etc.) but in the basic mechanisms through which that equipment has been produced.

Tracing pigment cells, muscle cells, or blood cells back into earlier phases of their development, we find, or postulate, that they contain then, instead of the terminal products of melanin, myosin, or hemoglobin, half-finished "precursors." But if we go back still further, beyond the "precursors of precursors," the future products lose their identity and we are faced with nothing but the plant capable of turning them out, and perhaps some of the ingredients. Now, evidently a plant designed to turn out chiefly melanin must be different from one designed to make contractile myofibrils. We express this fact in the names "melanoblast" and "myoblast" given to those early stages, and we generally concede that they are intrinsically different, although the differences are no longer demonstrable by ordinary microscopic or chemical techniques. By further refinements of technique we can extend the range of our discriminative powers; structurally, by the ultramicroscope, polarization microscope, electron microscope and x-ray diffraction; chemically, by microtests (including cytological stains), spectroscopy, microincineration, etc. These techniques will undoubtedly reveal criteria of differentiation much earlier in the embryonic history than was possible with the cruder classical tools of observation. This is a line of investigation worthy of vigorous prosecution. But we must expect that it too will come to an end short of retracing the whole ontogenetic course of a given cell strain, for we are still essentially concerned with tracing "products" rather than the production processes that lie behind. Once we can detect the "product," most of the story of its production is over and we have certainly missed its essential beginnings. It looks as if the analytical methods at hand are of no avail.

Preformistic theory would deny the validity of our dilemma. It would assert that all distinctions of later stages have existed, though less manifestly, from the beginnings of ontogeny. But experimental embryology has thoroughly discredited this concept by proving that mature cells which differ profoundly in their physico-chemical constitution may be derived from cells that are demonstrably equivalent and, to all practical purposes, identical. When and by what means have their developmental courses then become divergent? This is the crucial question of cytodifferentiation. Fortunately, though analytical techniques fail us, we can answer it by a test that might be called "behavioral." This is where we must draw on logics rather than gadgets.

Suppose we are to compare two closed material systems, whether they be molecules or cells or organisms, about whose character and composition we know nothing. How are we to tell whether they are the same or different? The only way of telling is by watching their behavior. If both behave identically under identical conditions, we decide that they are alike. However, if both, again *under identical conditions, behave differently*, then we must conclude that *they have been intrinsically different* from the very beginning of the test.

Against the cogency of this postulate even the most persistent failures to detect more tangible signs of difference would prevail nothing. Two seeds giving rise to different plants in the same soil, or two eggs giving rise to different animals in the same pond, would have to be acknowledged as intrinsically different even if they were otherwise indistinguishable. Similarly, if two cells under identical conditions behave and respond differently in any respect (e.g. movement, growth rate, chemical activity, sensitivity to radiation or drugs, etc.), then they *are* different, which implies that if they have descended from a common source of identical cells, they must have *become* different, i.e. have gone through a process of differentiation, no matter whether they show it by morphological signs or not. The behavioral test is as compelling as any morphological test, and far more pertinent; for morphological criteria are but residues and convenient indicators of prior activities.

Early differences of this type, which are not immediately recognizable and can only be deduced from later formative activities, are commonly described as differences of "potency." If this were to imply that they are virtual rather than real (in the sense of Driesch), then it would mean reinstating the old "vital spirits" under another name. Actually there can be no doubt that we are dealing with differences of equipment and conditions which are real, yet too subtle to be detectable by present techniques.

In conclusion, we propose to define "differentiation" as the unidirectional changes in the character of cells and cell generations during their life history, transformations by which they become increasingly different from their own former selves, their parent cells, and the cells of other strains which have taken divergent courses. And, as we have already intimated, "character" is to include the totality of reactions of which the cell is capable at a given stage. In any given situation, only a fraction of these possible reactions will be realized, and only a small part of this fraction will become externally recognizable.

Modulation

We have said that cells that look alike may be intrinsically very different. The opposite is equally true, namely, cells may look very different and yet be essentially alike in character. Let me cite a striking example from my own experience, the vagaries of the adult sheath cell of Schwann of adult mammalian nerve. In the embryonic stage these cells are long spindles. They then envelop the nerve axons as thin protoplasmic cylinders. When axons degenerate, as after nerve section, the sheath cells regain mass, become spindly again, and glide out from the nerve tubes. After nerve regeneration they resume their original sheathing position and corresponding cylindrical shape. But they can vary even much more than just shuttling back and forth between these two states.

Embryonic Schwann cells, reared in tissue culture along the interface of cover glass and fluid medium, tend to transform from the spindly type into a flat round form of the appearance and behavior of macrophages (49). A repetition of the experiments with adult nerve proved that the fully mature Schwann cell is capable of the same transformations and more (58). Plate I illustrates a series of cell forms obtained in vitro from Schwann cells of peripheral nerve.[2] They include tandem strands of filamentous sheath cells, ramified astrocyte-like forms, giant macrophages with ruffled membrane in active phagocytosis, monocyte-like round cells with clear ectoplasm and eccentric nucleus, and foamy round cells whose vitality is proven by their mitotic activity. Each one of these phases occurs under certain definable conditions, among which the fine-structural constitution of the medium seems to play a predominating role. Some of the transformations are easily reversible, others have as yet been observed in one direction only. Since they can be followed directly as they take place in the cultures, valuable

[2] Save for a brief note, the results of these experiments have not yet been published.

insight into the processes underlying the morphological results has been obtained. There was evidence of significant changes in the distribution of the cell content, size of the ectoplasmic border, constitution, elasticity, rigidity, mobility, etc., of the cell surface, average number and shape of pseudopodia, mode and rate of locomotion, viscosity, vacuolization, adhesiveness, orientation of fibrillar plasma constituents (as indicated by silver impregnation), fat production, distribution of nuclear materials, shape and hydration of nucleus and nucleoli, phagocytic power, and other associated properties. In short, these cells, in response to different environments, have undergone rather profound reorganizations of the kind unhesitatingly classified as "differentiation" or "metaplasia" by a nomenclature based on sheer appearances.

But is this type of change directly comparable to the progressive cytodifferentiation during ontogeny of which we have spoken before? It evidently is not, and it is in this place that facts supplement logic.

As we mentioned before, many of the observed changes are reversible if the cell is returned to its prior environment. There are many more cell types of the mature organism which behave similarly in that their behavior and shape change reversibly with their environment. Changes in the nutrient and hormone composition of the internal milieu, sublethal toxic agents, inflammatory processes, seasonal fluctuations, denervation, and in fact, on a more moderate scale, practically all ordinary physiological stimulations, cause a more or less thorough reshuffling of the cell content with consequent change in cell behavior, and often in cell form. The important point to keep in mind is that in all these instances the cell can revert to its original state when the original environmental conditions are restored. Evidently the cell passes through these transmutations with no alteration, either gain or loss, of its basic equipment. At the end of a cyclic change it turns out to be the same cell that it was before, and it is quite beside the point that sometimes the reversal may not occur in the same cell, but in one of its descendants which then prove to be "chips off the old block." There is no reason why the reversible changes observed in our Schwann cells should be regarded in any different light than, let us say, the contractions and expansions of a melanophore in response to cyclic nervous or humoral stimuli, or the change in the secretory state (and associated morphology) of a hormone-sensitive cell in response to varying concentrations of the respective hormone. They all are merely expressions of how a given cell can react to a variety of conditions—its "reaction repertory," based on its material constitution such as it is at the time. The living cell is an unstable system which can have neither existence nor form except in relation to its environment. Few animal cells (e.g. red blood cells) freeze their shape by assuming a rigid envelope. For the rest, shape is variable, merely an index of substance disposition attained in given surroundings and valid only for these particular surroundings.

Obviously, then, the misleading mental picture of absoluteness and static rigidity of cell shape, conjured by the microscopic picture of fixed cells on slides or in textbooks, must be replaced by a dynamic definition based not on the momentary expression of the cell but on all the possible reactions of which it is capable. The variety of expressions that thus can be assumed by the cell on any level of differentiation I have called "modulations" (4, 47). If we may draw a chemical analogy, differentiation would be comparable to the synthesis of a new compound by an irreversible reaction chain, while modulation would correspond to the isomerism and allotropy of intermediate or terminal products. The former involves a change of *composition*, the latter merely of *disposition* in space, of the constituent elements. The implications of this simile for the concept of the cell will be dealt with later.

It can be plainly recognized now how past concepts of differentiation have been vitiated by undue reliance on visual criteria. Cells that looked alike were thought to be essentially of like character, although they often were radically different; and conversely, cells that looked very different were considered as very disparate, although they actually were often of the same kind. The literature of histology, embryology, and pathology abounds with notable examples of victims of this "visual illusion."

Differentiation and Dedifferentiation

If the test of differentiation, in contradistinction to modulation, is the unidirectional, progressive, and irreversible trend of the former, how safe is the evidence that differentiation in this strict sense ever really occurs? Might not all differentiations be cases of modulation, merely obscured by the difficulty of always finding the right conditions to reverse the fate of the cell?

As long as we confine ourselves to normal development, where different cell behavior is generally associated with different local conditions, the answer remains in doubt. For a crucial test of whether different cells differ in character, we must remove them from their different environments and observe them (*a*) in identical environments, and (*b*) in each other's places. This can be done by the techniques of explantation (tissue culture) and of transplantation, respectively. The answer has been unequivocal.

Cells taken from different tissues of late embryonic or mature organisms and transferred into standard extraneous media offering nutrition, protection, and support lose many of their specialized aspects, i.e. criteria by which we used to tell their "differentiation." Spectacular as their differences may have been during residence in the body, they soon come to look and behave much more alike, in adaptation to the common environment. In the early days of tissue culture this simplification of appearance was actually often interpreted as a sign of "dedifferentiation," in the sense of a return not only to a more primitive but to a veritably primordial state. This was again judgment by appearances with all its pitfalls. Yet later work showed clearly that the explanted cells, in spite of their similar disguise, retain distinguishing features of fundamental nature which they pass on unaltered to their descendant cells for indefinite numbers of generations. As a glance at Plate I will illustrate, Schwann cells and endoneurial connective tissue cells, while assuming both the aspect of common "mesenchym," i.e. spindle shape and amoeboid activity, can be well distinguished by their sizes, nuclear characteristics, staining properties, affiliations, and other criteria revealed only by subtler analysis.

In other words, the parallel modifications which the cell strains undergo on transfer to tissue culture are merely modulations carried out by variously "differentiated" cells that otherwise preserve their basic distinctions. Their distinctiveness persists even during such further modulations in culture as the transformation to the macrophage stage. This transformation has thus far been observed in cultures of various connective tissues (25), Schwann cells (see above), and muscle (7). They all can give rise under similar conditions to a highly phagocytic amoeboid cell with hydrated nucleus, dense entoplasm, and a more hyaline ruffled border, i.e. macrophage-like characters. But in doing so they also retain the signs of their descendancy, so that a "macrophage" resulting from the conversion of an endoneurial cell can usually be clearly distinguished from one that has come from a Schwann cell.

Tissue culture thus does not really abolish character differences established during ontogeny. This conclusion is fully borne out by the critical literature on the subject (4, 40). It is strikingly confirmed by experiments in which cells that had lost their more specialized aspects during intensive proliferation in tissue culture were again given opportunity to produce their specialties, either by suppressing further proliferation in vitro or by transplanting them back into an organism. Under such conditions they were essentially true to the kind from which they stemmed: those derived from pigment epithelium would resume the production of pigment granules (8); those from thyroid, of colloid (13); those from gut, of digestive enzymes (14). Cultured cells returned to an organism, but into atypical locations (42), moreover, showed no tendency to differentiate in conformity with their new environment, proving that the culture period had not restored to them the ability of younger cells to switch into a new local course of differentiation.

All this adds up to the realization that the measure of true differentiation is the absence of true dedifferentiation, if dedifferentiation connotes not the mere assumption of a semblance of primitivity but the actual return to a state of wider potency, i.e. the recuperation of something that had ostensibly been lost. By the testimony of the reported and similar experiments, vertebrate cells are incapable of such true dedifferentiation; hence, for them, true differentiation is the rule.

Tissue Differentiation

All our statements thus far refer to cytodifferentiation. Histodifferentiation and organ differentiation require some additional comment. Tissues and organs consist of cells and matrix in definite mutual relations and groupings. What happens during the differentiation of a tissue is in large measure a direct reflection of the differentiation of its constituent cells. Their order of proliferation, their movements, transformations, affinities, secretions, etc., and the spatial arrangements and restraining conditions resulting from group occupancy of a common space determine, for the main part, the character and organization of the tissue or organ. By their collective activity the cells set up conditions for one another that would not develop if they remained single and independent. For this reason, what is commonly called "histodifferentiation" implies an added element of order and novelty over the mere "cytodifferentiation" of the unit cells. This is particularly striking in the formation of composite organs where the combination of cells from diverse sources leads to novel results which neither of the contributing components could have produced by itself. In addition to material contributions, there are dynamic influences of mechanical, chemical, and electrical nature acting from one tissue or organ upon another.

But on close inspection there remains no property of a tissue that could not be ultimately traced to the activity of either its own or other cells. It is by virtue of properties acquired in cytodifferentiation that cells can combine, interact, and arrange themselves in certain specific ways, conditions permitting. "Histodifferentiation" implies merely the existence of conditions under which cytodifferentiated cells can realize their individual and group faculties. This helps to clarify the meaning of the term "tissue dedifferentiation." Evidently cells may lose their typical associations and orderly arrangements just as we have seen before that they may lose internal differentiation products, without giving up their essential character. Destruction of intercellular matrix, for instance, can set cells free, and such cells may resume motility, scatter, and modulate in various ways. Their community has been dissolved, but they themselves have retained their identity. To this extent, what is called "tissue dedifferentiation" is merely tissue disintegration. Since mobilization of cells is frequently attended by the dissolution of internal differentiation products (e.g. myofibrils), the superficial picture becomes one of general simplification and regression. There is no harm in continuing to speak of tissue dedifferentiation, so long as one bears in mind that it need involve no true dedifferentiation of cells.

Which Cell Parts Differentiate?

A cell is a highly complex and heterogeneous system. Then where specifically do the changes occur that constitute differentiation? In the cytoplasm, in the nucleus, or in both? Or rather, which ones of their subdivisions undergo change, and which do not?

It is customary in speaking of cytodifferentiation to emphasize mainly the concomitant profound changes in the content of the cytoplasm. But the nucleus likewise undergoes patent modifications, which become evident in the divergence of nuclear characters and behavior among different cell types. Some of these nuclear distinctions can be deduced from differential reactions to histological stains, which in essence constitute microchemical tests. In special instances even more direct signs of nuclear differentiation have been observed, as in the extrusion of formed secretion bodies from the nuclei of certain nerve cells (30), or in the transfer of the major mineral content from nucleus to cytoplasm during the maturation of neuroblasts (36). On the other hand, it is a basic tenet of genetics that the chromosomes, or at least those parts of them that constitute the genes, remain unaltered in the process of somatic differentiation. Accordingly each nucleus contains both "differentiated" and "undifferentiated" portions.

The question arises whether the cytoplasm might not be similarly subdivided into a differentiated fraction and an undifferentiated residue. Such is generally conceded to be the case for the cytoplasm in the "germ track," i.e. of those cells that lie in direct ascendancy of the primordial germ cells. The same must be assumed for those "totipotent" somatic cells of the lower invertebrates which can give rise to a new individual by budding, regeneration, or other forms of asexual reproduction. The specialized somatic cells of higher forms, on the other hand, have never given any evidence of such totipotence. This could be interpreted to indicate that *all* of their cytoplasm has been diverted into a specialized course, has become liver-specific for the liver cell, kidney-specific for the kidney cell, and so forth. Yet again, there is undeniably some stock of cytoplasmic equipment that practically all cells have in common (e.g. respiratory enzymes, mitochondria, centrosomes, spindle fibers, etc.) and which, thus, seems to have escaped the progressive changes of cytodifferentiation.

In short, any attempts at making general statements for the nucleus as a whole or the cytoplasm as a whole end up in contradictions. The more concretely and specifically we try to state our problems, the more clearly we realize the inadequacy and ambiguity of our traditional vocabulary. So long as we address our questions not to a real living cell, but to a vague and abstract symbol, we can expect no more than oracular replies. Those general concepts of cell and cytoplasm with which we conventionally operate, with their sundry textbook attributes of contractility, excitability, adaptability, and so forth, have just about lived out the term of their usefulness. We no longer view contractility and excitability vaguely as "fundamental attributes of protoplasm," but have recognized them as defined properties of circumscribed physicochemical systems, protein chains and surface membranes, respectively. Differentiation deserves a similarly specific treatment. This will be feasible only if we replace the abstract notion of the cell by a much more specific picture. Such a picture is gradually emerging from the combined efforts of cytology, general physiology, and biochemistry. According to this picture, a cell is made up of known or knowable molecular elements, and cellular activities are resolvable into elementary molecular processes.

Study of these elementary processes is well under way and has yielded spectacular results. However, preoccupation with the elements has also at times tended to obscure the fact that in order to constitute a living cell the constituent molecular processes must not just happen, but must take place according to a highly ordered plan of interrelations based on orderly spatial distributions, orderly chronological sequences, and orderly ratios of rates. No model of a cell can be pertinent unless it takes into account both the elementary processes and their organizational frame. It is imperative, therefore, that we match research on the former by an equally vigorous search for the physical basis of the latter. To facilitate this task I have tried to sketch a sort of crude molecular blueprint of the cell, in which *the molecular phenomena are placed in relation to the organization of the cell as a unit.* This picture is of necessity diagrammatic, oversimplified, tentative, and full of blank spaces, but it does serve to crystallize the problems. Most of it has already been presented on an earlier occasion (54).

Molecular Ecology

Science advances not only by discovering new truths but also by eliminating old errors. Most older cell concepts held to the view that cellular organization is based on rigid structural frameworks of microscopic order, but this view has been ruled out by modern cell research. Cytoplasm behaves as a viscous liquid. It can be thoroughly mixed up by stirring or centrifugation, and yet its countless chemical workshops will continue to operate in good order, which implies proper spatial segregation. No orderly microscopic array could have survived in the commingled content; what has survived is a fixed set of conditions—the rule, as it were—according to which the original order can be restored. And where do those conditions reside? Evidently they must be looked for in those parts of the system which can preserve organization in the shuffle. There are two categories of cell constituents that qualify under this title, and they are of quite different order of magnitude, one molecular and the other supramolecular. The former consists of the large organic molecules or molecular aggregates with their specific structure and faculty to perpetuate this structure in the presence of the necessary ingredients and appropriate sources of metabolic energy. The latter consists of the existing surfaces and interfaces of microscopic or submicroscopic order, whose molecular linings are anchored by adsorption or chemical fixation and thus better protected against random disruption.

Organization within any sample space of a cell is then to be regarded as a result of the counterplay between these two poles: the specific faculties for selective combination and segregation of the molecular constituents on the one hand; and the marginal reference frames of surfaces on the other, which, by their power of adsorbing and orienting the mobile elements, force them into a supraelemental collective order. It seems impossible to account for cellular organization otherwise than by such a dual concept. Its essence can be summarized as follows: Manifest cell organization results from the response of organized elements to fields of organized (i.e. non-random) physical and chemical conditions, here tentatively identified with conditions prevailing along interfaces.

The nearest simile to this type of organization is found in the behavior of human or animal populations, which likewise consists of the reaction of organized individuals to an organized environment. In both cases, the order is one of *ecological* conditions. Thus, cell biology, in molecular terms, becomes not merely molecular physics and chemistry of the cellular constituents, but *molecular ecology* (54), a science not of molecular individuals or pairs, but of *molecular populations and their existential conditions.*

Some day we may hope to have a fairly complete list of the chemical compounds present in any given cell, from simple elements to the highly complex organic macromolecules, and thus develop a taxonomic catalogue of molecular species. But just as a museum collection of animals is never lifelike unless it shows them in their natural environment and mutual ecological relations, so this chemical catalogue would not be representative of the cell unless it were supplemented by information on the groupings, behavior, interdependence, and general operating conditions that tie the various molecular species into a durable, and indeed "viable," community.

Biochemistry is producing an ever-growing wealth of evidence for the complexity of molecular interdependence in living systems. Even very elementary physiological processes, as for instance the energy transfer in respiration or photosynthesis, involve a great number of coordinated steps, each of which requires the presence of a definite set of properly dosed reactants, mediators (enzymes), and physical conditions (*p*H, temperature, pressure). Undisturbed operation of these energy-yielding systems, in turn, is a prerequisite for the synthesis, among others, of the complex protein molecules, some of which, as enzymes, are again fed back as indispensable links into the energy-producing systems. Thus neither system can maintain itself without the other. Protein synthesis itself proceeds in series of steps, each of which can occur only under specified conditions which will vary for different compounds. The various chemical systems of the cell thus form "symbiotic" interdependences, and the integrity of the cellular system as a whole is contingent on their coexistence and cooperation. For each particular chemical system we can thus postulate a definite set of "existential and operational prerequisites," that is, conditions indispensable for the synthesis, preservation, and operation of specific organized products. The character of a given animal population is determined by the physical environment (climate, physiographic factors, etc.) and by the organic environment (food, competition among species, biocoenosis, etc.) in that particular area. Similarly, the character of the molecular population occupying a given site in a cell will depend on the physical conditions (electric potentials, surface tensions, etc.) and the chemical interactions among the various molecular species in that locality.

The ecological simile is primarily a convenient device to describe the behavior of the molecular populations of cells. It permits us, for instance, to answer in principle the question of how the morphological segregation and orderly distribution of different chemical systems in the cell is brought about, preserved, and restored after disturbance, despite the absence of static internal frameworks. If each complex biochemical system requires highly specific conditions for its formation and operation, then obviously the probability of finding that kind of system will be high at sites where those conditions are satisfied, and low at others. And after commingling, the content will again sort itself out according to the frame of marginal conditions, and if the latter have remained undisturbed, the re-sorting will likewise follow the old order. We will thus find preferential sites for the synthesis of certain compounds, namely, sites where the necessary ingredients, templates, energy sources, and physical factors for that particular synthesis coexist. There must also be preferential sites for the settlement of the finished products; for as synthesis continues, the accumulating products are driven further and further from their production centers and thus brought under conditions not necessarily compatible with their continued existence. I have suggested on a previous occasion (54) that such preferential settlement might be a matter of the configuration of the molecule, as a stable layer of molecules with specifically shaped ends firmly adsorbed to an interface could readily trap roaming molecules of complementarily fitting configuration.[3]

[3] This hypothesis leans on the immunochemical concepts of Pauling (32).

The concept of molecular ecology gives a more truly representative picture of cell life than the older morphological concepts. It gives expression to the fact that what is "determined" in a given cell is only the general statistical norm of the proportions, arrangement, and distribution of its constituent elements, but not an absolutely fixed and stereotyped pattern exactly repeated in each individual cell specimen. No two cells are ever precisely congruous. It is in the very nature of the cell that we can predict no more than the degree of *probability* of the occurrence of a particular process in a given place, and this only with reference to what is happening in the rest of the cell.

In animal ecology the sizes of the mobile units vary greatly. Some animals move independently, others in flocks, and still others as heterogeneous groups. Similarly, in the molecular populations of the cell some molecules move singly, others combine into larger bodies, submicroscopic or microscopic particulates, and still others form mixed aggregates, with members of different species compounded in definite proportions (e.g. lipoprotein complexes, coacervates, etc.). The formation of any such larger unit introduces a new element of complexity (new specific surface, new conditions for adsorption, restraints of mobility, etc.), and if one thinks the matter through to its logical conclusion, one realizes that cytodifferentiation effects its manifest order through a series of steps of progressive complication, in which the basic determinants are the properties and mutual affinities of the different molecular species, and a frame of "conditions" prevailing in their common space, favoring the various species differentially. This frame or ground plan, in relation to which the mobile elements will become ordered and sorted, must evidently possess a greater measure of physical stability than the rest of the system. In a relatively fluid system, such as a cell, the entities to answer this demand are the surfaces, in which molecular mobility is restrained. Surfaces thus assume foremost rank as ecological niches for the assembly and segregation of different segments of the molecular population of the cell. Their mode of operation can be briefly sketched as follows.

Surface Organization

Let us consider the interfaces which set off a cell from its surroundings, or a cytoplasmic inclusion from the matrix, or a chromosome from the nuclear medium. Heterogeneous systems meeting along a common border immediately affect each other across the border. This is due to the fact that the molecular population of one system begins to react differentially to the conditions introduced by the new contact system (electric charges, surface tensions, combinations and reactions between border molecules). The resulting interactions lead to (1) the segregation of a selected fraction of the molecular population into a surface layer, and (2) the immobilization, hence, solidification, of this portion. The mechanisms involved will vary, but for our present purpose they can simply be lumped under a common term as "*selective adsorption.*" Their common denominator is the fact that a given interface of given constitution will intercept and retain certain member species of a mixed molecular population more easily and more strongly than others. Consequently, the favored species will become concentrated at the surface and the less favored ones correspondingly diluted. A change in the outer conditions may force the replacement of the former border population by new and more appropriate species from the interior. A complex system of this kind will have a labile phase, during which the molecular surface grouping responds to changes in the physical state and chemical composition of the adjacent outer system, and a secondary stable phase, which follows the consolidation of the surface due to chemical bonding, gelation, precipitation, etc. If the outer conditions are not uniform over the entire surface, but vary locally, the adsorbed molecular populations will be correspondingly variegated. The same cell can thus develop different types of surface zones on different sides exposed to different environments. Moreover, each of the manifold types of interfaces inside the cell will also acquire its own appropriate type of covering. The segregative power of specific surfaces can thus be readily understood.

In becoming adsorbed to an interface, polar molecules assume a common orientation. Their exposed ends thus constitute a new free surface to which further molecular species of fitting configuration may be built on in a second layer. This, in turn, may serve for the deposition of a third group, and so layer upon layer may be stacked up in a specific sequence, the molecules of each interlocking with those of the next.[4] Although factual information is scarce, it might be surmised that the apposition of new layers in living systems is governed by highly specific relations among the combining molecules, perhaps due to the steric fitting, key-lock fashion, of the specifically shaped ends of the conjugat-

[4] Such stacking up of molecular films has actually been demonstrated by Langmuir and Blodgett (23), and although their observations were made on rather simple models, evidence of a similar molecular lamination at the surface has been found in at least certain types of cells (10, 34).

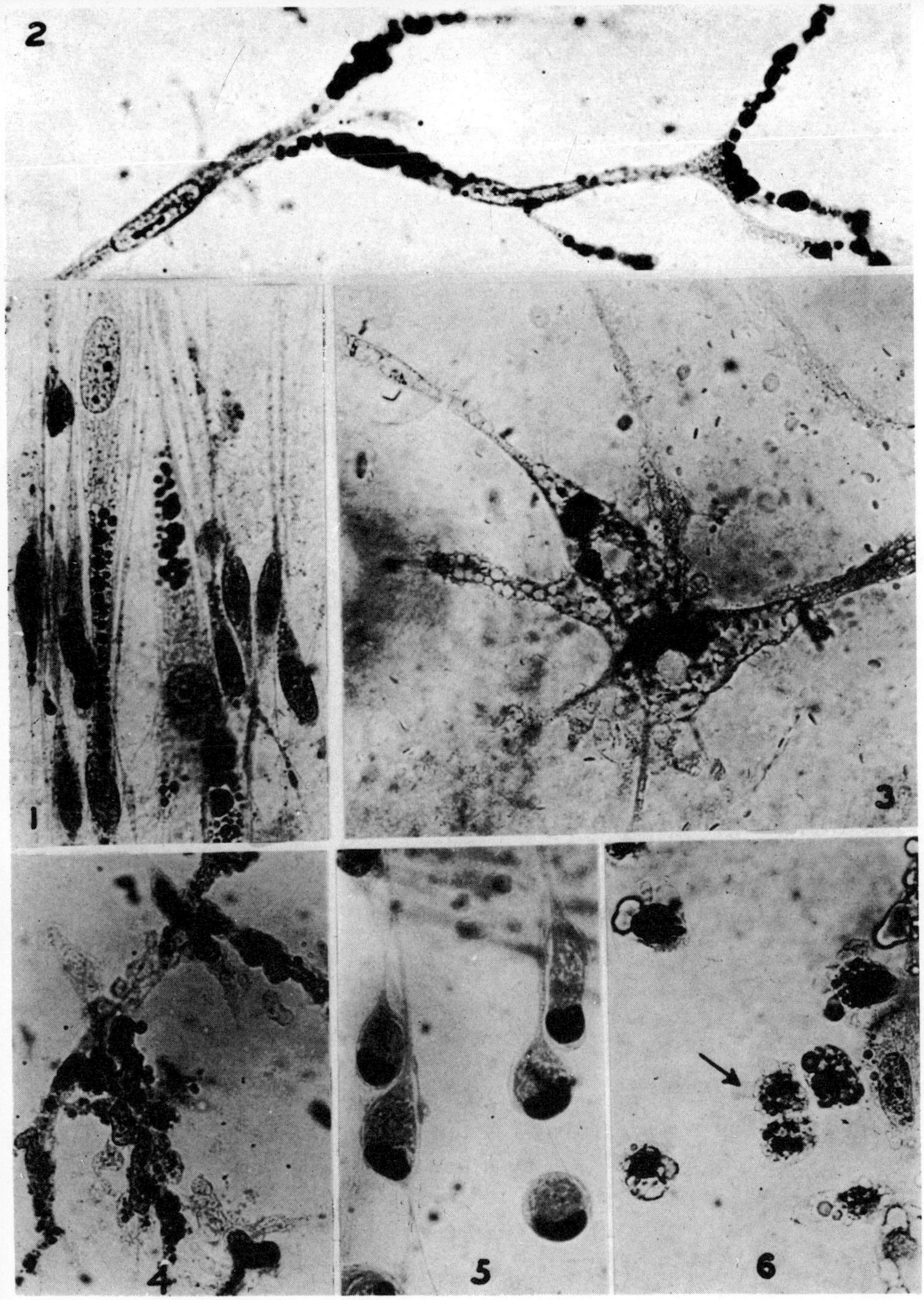

PLATE I. Transformations of Schwann cells from peripheral nerves of adult rats in tissue culture (modified from Weiss and Wang, 1945). All figures at identical magnification (× 710). *Panel 1* shows the original slender spindle shape of Schwann cells freshly emigrating from nerve. The picture also contains two large endoneurial fibroblasts (large, oval, lightly-stained nuclei; dark, sudan-stained fat droplets in cytoplasm). *Panel 2* shows beginning ramification of Schwann cells in plasma clot (note rod-shaped nuclei and fat droplets). *Panel 3* shows giant multinucleate "macrophage" developed through transformation and coalescence of a group of Schwann cells at interphase between cover glass and capillary exudate below plasma clot. *Panel 4* shows extensively arborized Schwann cells with polymorphic nuclei at fringe of culture (note active extrusion of fat bodies). *Panel 5* shows transformation stages from spindle type through monopolar "flask" shape to spherical "monocytoid" form with sickle-shaped eccentric nucleus and distinct zone of hyaline exoplasm. *Panel 6* shows small macrophages formed from Schwann cells in liquid areas of medium. (Note mitosis in telophase indicated by arrow.)

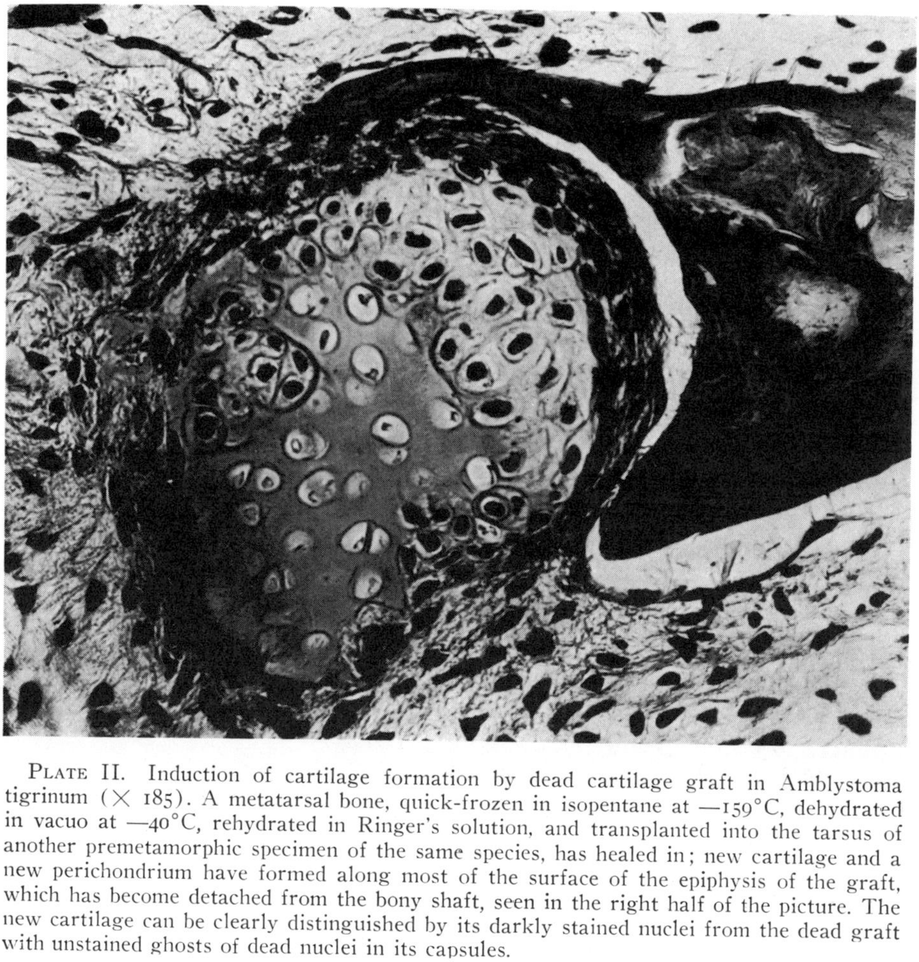

PLATE II. Induction of cartilage formation by dead cartilage graft in Amblystoma tigrinum (× 185). A metatarsal bone, quick-frozen in isopentane at —159°C, dehydrated in vacuo at —40°C, rehydrated in Ringer's solution, and transplanted into the tarsus of another premetamorphic specimen of the same species, has healed in; new cartilage and a new perichondrium have formed along most of the surface of the epiphysis of the graft, which has become detached from the bony shaft, seen in the right half of the picture. The new cartilage can be clearly distinguished by its darkly stained nuclei from the dead graft with unstained ghosts of dead nuclei in its capsules.

ing molecules (54). As complex macromolecules will mostly have different configurations at opposite ends, successive layers will consist of different molecular species. Here we have, therefore, a process that, once it has started from a given surface, could lead to a progressive segregation with stratification of molecular species which were originally intermingled. In this process molecular orientation is of paramount importance, since it exposes the maximum number of specific receptive ends to the interior. This consideration gains added significance in the case of those polar molecules that have enzymatic properties, as parallel orientation and packing might increase their effectiveness by turning all active groups toward the substrate. Then selective surface adsorption would become a prime factor not only in the segregation of existing molecular species, but also in the creation of new ones. Moreover, the molecular fringe which settles along the surface acquires master control over substance transfer across the cell boundary in that it can selectively facilitate or prevent the exchange of materials between the areas it divides.

The extent to which a surface layer grows inward by apposition, will vary with the circumstances. For example, in some cells the outermost layers form a thick and distinctive coat (myelin sheath of nerve fiber, envelope of red blood cell, vitelline membrane of egg), while in others they remain an inconspicuous "plasma membrane." The subjacent layers (crust of egg, ectoplasms) likewise vary in thickness as well as in sharpness of demarcation. It must be remembered, however, that stratified molecular segregation need not be so massive as to become microscopically discernible.

The picture of surface organization here presented is incomplete and greatly oversimplified. For instance the specific key molecules, which were given such prominence, obviously lie interspersed with many more, smaller, simpler and trivial molecules of less specific behavior, a mosaic feature which we have not duly taken into account. It has been suggested on the basis of studies on drug action, immunological reactions, and radiation effects, that target points of specific response are present in isolated patches occupying only a small fraction of the cell surface. A similar discontinuity characterizes the structure of the chromosomes. Accordingly, many of the specific effects assigned above to surfaces in general will actually be confined to certain portions of the total surface only. An extension of this thought leads to the realization that complex molecular aggregates (particulates) may be subject to specific adsorption and fixation, with one of their constituent key patches effecting the specific attachment, and the rest remaining uninvolved. But to discuss these and other qualifications in detail, would exceed the scope of this paper.

Differentiation in Molecular Terms

We return from this excursion into the molecular realm better equipped to phrase questions about differentiation and growth in tangible form. We can now ask, for instance, whether differentiation really signifies a change in the composition of the molecular population of the cell or merely a change in the distribution of preexisting molecular species. The latter view has been implicit in those embryological and pathological theories which hold that the various cells of a given organism are all composed of essentially the same basic and immutable "protoplasm," and that their diverse appearances and performances are merely the overt responses to a variety of external situations or "stimuli." A given response display (called "differentiation") is thought to last only as long as the given stimulus situation lasts. Change the latter and the display will change accordingly ("dedifferentiate" and "redifferentiate") in a new direction.[5] Yet, when we view the facts from the molecular level, there seems to be no doubt that cytodifferentiation does not leave the chemical composition of "protoplasm" basically unaltered.

The ubiquitous species of small and simple molecules going in and out of cells remain about the same at all stages except for differences in concentration. Some species of large and complex molecules, including many of the proteins, are also present throughout development (5, 28). All these comprise what we have called above the stock in common to all cells. But in addition to these, differentiation brings with it, or virtually consists of, the continuous elaboration of molecular novelties which have not been present from the beginning. Thus, a liver cell, a nerve cell, a pigment cell, a mucus cell, a thyroid cell, etc., each produces its own chemical specialties. Can this really be taken as evidence that the respective "protoplasms" are chemically different? Might it not be that all mature cells have the synthetic faculty for the whole array of cell products known to the body, and that a given cell is merely prevented by local external conditions from materializing any but the appropriate one? This contention is contradicted by the experimental evidence quoted above, showing that cells transferred into a different environment

[5] A certain partiality toward such a view can be detected, for example, in the earlier writings of Child, and there have appeared several more modern, if less thoughtful, versions of the same theme.

(explantation or transplantation), though they may cease to produce their customary products, never reacquire the faculty to give rise to any but the products of their original line, when given another chance.

Thus the conclusion is inescapable that the various cell types mentioned do not just temporarily exhibit different production processes, but have themselves become chemically converted into different production plants. That is, each cell type develops, in addition to the molecular species which it contains in common with all or several other cell types, its own peculiar and distinctive array of molecular species. These are the ones that account for the differential behavior of the various protoplasms; they are the true objects of differentiation. One is tempted to identify them with specific proteins, but factual knowledge is still far too inadequate for any generalization. We need much more precise information on the time and mode of appearance of various cytochemical constituents in the ontogeny of various cell strains. Such an inventory is bound to contain the answers to numerous questions about which we can as yet do no more than speculate.

In contrast to differentiation, cell modulation implies no essential change in the composition of the molecular population, but merely a regrouping of the existing species. According to our previous discussion, any change in the environment of the cell that alters the conditions in the surface could cause a reshuffling of the content and replacement of the surface population with marked changes in the behavior and morphology of the cell, but without necessarily affecting its basic composition. However, whether such a modulation is fully reversible, i.e. whether or not the reshuffled population can return once more to its initial distribution when the original environmental conditions are restored, depends on the degree of consolidation the surface has undergone in the interim.

These considerations suggest the possibility that all differentiations start out as modulation, as mere molecular regrouping, in response to changes in the cellular environment. Let us consider an embryonic cell shifting from a site *A* to another site *B*, or, what amounts to the same, a change in the environment of a stationary cell from a condition *A* to a condition *B*. Exposure to the new contact conditions brings molecules with affinities to *B* to the surface. Now, suppose this molecular coat initiates a chain of chemical reactions, through which the composition of the interior is materially changed. This implies the appearance of novel molecular types as well as the disappearance of some of the existing species as they become converted into new ones. Such change cannot take place all at once throughout the cell. We must assume that it starts from a localized area (in our concept, the surface) and spreads at a measurable rate. Therefore if the cell were returned to its original condition *A* during the early phases of this process, it would still contain sufficient amounts of all the species it had before to resume its original state. To the observer this would appear as reversible modulation. But continued residence under condition *B* would permit continued progress of the changes indicated, until eventually the cell will have lost a major segment of its earlier constituents. It would thus have undergone an irreversible change of its constitution, i.e. true differentiation. The lability of differentiation in its incipient stages is a matter of record.

If a cell that has assumed character *B* then moves on to a site *C* (or if its neighbor cells or the intercellular milieu changes to a condition *C*), its surface population will again become reconditioned and then initiate a new chain of processes, marking a further step of differentiation, the earliest phase of which would again be labile. A cell transferred from *A* to *C* directly may or may not respond in the same fashion as when coming by way of *B*, depending on how decisively its molecular population has been altered during the intermediary stay at *B*.

We are portraying "differentiation" as a chain of events consisting of alternate physical regrouping and chemical alteration of the molecular populations, the latter phenomenon involving the emergence of novel species of compounds. It can be seen that according to this concept the potentialities for differentiation are strictly limited by the initial chemical endowment of the cell and that they become further restricted as the cell passes through the various phases *A, B, C*. What we conventionally call "loss of potency" is therefore merely the counterpart of the positive increase in definite chemical specialization incurred by the "differentiating" segments of the molecular population. We need assume no separate inhibitory agents.

This concept contains also, in principle, the answer to the problem of divergent cytodifferentiation, that is, of how two initially identical cells may become the source of two qualitatively different strains. Let us assume that of two cells with identical molecular populations, including key species α, β, γ, δ, ϵ, etc., one is exposed to an environment *E*, and the other to a different environment *F*. Given some agitation and sufficient mobility of the cell contents, we may further assume that in each cell a selected fraction of key species will be concentrated at the surface, namely, those species which best conform to the adjacent medium; let us say, compounds ϵ in the cell exposed to the E surface, and compounds ϕ in that at the F surface, where ϵ and ϕ may be entirely unrelated. According to the preceding section, such adsorbed and oriented border species can exert key effects in turning the chemical reactions, segregations, and syntheses among the rest of the population into a definite course. Consequently the chemical fates of the two cells, coated by qualitatively different key species, will likewise become qualitatively different. Thus a difference which started out as merely one of distribution gradually develops into one of composition and character. This would be the fundamental step in the dichotomy of cell fates. This concept remains, for the present, hypothetical. But being more concrete and specific than the conventional verbal concepts, it has the advantage over the latter of being amenable to experimental verification. Tentatively, it fits the known facts of development very satisfactorily, as will be shown in the following.

On an earlier occasion (47) I outlined three basic principles of cytodifferentiation, which I called (1) discreteness, (2) exclusivity, and (3) genetic limitation. Any valid concept of differentiation must be consistent with these principles.

The principle of discreteness refers to the fact that "differentiation produces a definite number of discrete, distinct, discontinuous and more or less sharply delimited cell types which are not connected by intergradations." The principle of exclusivity states that once one type of differentiation has taken hold of a cell, no fraction of such a cell can ever become engaged concurrently in any other type of differentiation; the cell in differentiation acts as an entity in all-or-none fashion. For instance, there is a sharp dividing line at the rim of the optic cup separating optic retina from tapetum; one cell is a typical retinal cell and the next cell is just as typical a pigmented tapetum cell. Although we know from experimental evidence that originally both have the endowment to transform into either type, they never form a half-way blend. Examples of this kind could be multiplied at will. It is evident that these principles, derived from empirical facts, can be logically deduced from the postulated discreteness of the molecular key species α, β, γ, δ, ϵ, and their mutual exclusiveness in the competition for surface positions. Thus, a given cell must follow either the α-determined or the β-determined or the γ-determined, etc., course, and follow it wholly and with no compromises.

The third principle, genetic limitation, stating that each cytodifferentiation occurs according to the special methods characteristic of the parent species, reflects the fact that the numbers and kinds of molecular key species available for differentiation are limited by the inherited, hence, genetically controlled, initial endowment. It would seem premature, however, to speculate about the relation between our postulated key species and genes.

Phenomena of dominance and recessivity of cellular activities can likewise be explained in terms of our hypothesis. Any molecular species assuming a monopolistic surface position thereby gains "dominance" over competitively inferior species which, being barred from the surface, are in no position to express themselves effectively.

INDUCTION

After this consideration of cellular responses in differentiation, some comment should be given to the factors calling them forth in the organized play of development. In view of the fact that the problems of "organization" are discussed by Dr. Nicholas in the next chapter, we shall forego any but the most cursory treatment of the subject here.

According to the testimony of experimental embryology, development proceeds as follows. Cleavage subdivides the egg into blastomeres, each of which receives in addition to a nucleus a certain fraction of the original cytoplasmic system. While the nuclei are essentially equivalent, the cytoplasmic allotments vary, depending on the degree of specific molecular segregation attained in the undivided egg. The role of the most nearly stabilized part of the egg, namely its surface, must be especially taken into account.[6] Different blastomeres inherit different parts of the original egg surface. This initial differential among egg parts is then further elaborated and amplified by the subsequent reactive transformations which the various cleaved-off cells undergo in response to the organized conditions of the system which they constitute.

Descriptively, these conditions have been referred to as "fields"; but we shall not concern ourselves here with their character and operation. We need only keep in mind that in the early phases of development any area of the cleaved egg is endowed with dual capacities. On the one hand, it is composed of a large number of individual cells which in many cases, according to experimental evidence, are practically equivalent ("equipotential") and hence interchangeable as far as their response faculties are concerned. On the other hand, the totality of these cells, as a collec-

[6] The organizational role of the surface of the egg has been emphasized by several embryologists (9, 21, 33).

tive unit, exercises specific functions not exhibited by the individual elements as such. To mention a familiar though crude example, the differential exposure to environmental factors of cells in the core of a given group, as contrasted with those near the surface, is a property of the collective not owned by its members.

Cell groups under the influence of such collective factors, but prior to the appearance of a manifest response on their part, are usually referred to as "determined." There follow material transformations of the kind discussed in the preceding chapter, introducing novel relations among parts. Stepwise, in assembly-line fashion, the diversity of activities increases. As initial time differences become accentuated, some parts of the germ soon outstrip others in their tempo of differentiation. From then on, influences of the more advanced ("determined") upon the less advanced ("undetermined") parts become increasingly prominent. These influences find essentially two forms of expression: (1) influences spreading within a primary cell continuum, and (2) influences exerted from one cell continuum upon another contiguous one, usually after secondary junction.

Influences of this type are customarily referred to as "inductions." The term is purely descriptive and covers a considerable variety of heterogeneous phenomena which have often no more in common than the label. There is not the least justification for the supposition that a single operative mechanism underlies them all, much less so a single chemical entity. Much more pertinent than the illusory search for a master solution to the "induction" problem is an objective analysis of each and every one of the various concrete instances of induction that have thus far been recognized. Therefore, instead of discussing induction in general, let us concentrate on two specific examples, one for each of the categories mentioned above.

The first example concerns what might be called "homoeo-induction," that is, the progressive recruiting of cells for a given type of differentiation, spreading from a focal area like an infectious wave (e.g. the antero-posterior progression of myotome formation). One gets the impression that cells which have attained a certain differentiated character can communicate their state to their neighbors, which then pass it on further, and so on down the line. The experimental test lies in transplanting a differentiated fragment amidst less far advanced cells (37). In many embryonic operations of this type, completion of the fragmentary graft structures by recruitment from surrounding cell sources has been observed, but in no case has the precise mechanism been revealed.

Technically it might be easier to start with post-embryonic stages, particularly regeneration. For instance, the fact that transplants of mature cartilage may induce new cartilage formation lends itself to a much more detailed analysis than has as yet been undertaken. Nageotte (27) reported that the implantation of alcohol-fixed cartilage into the rabbit's ear stimulates the formation of new cartilage in the adjacent connective tissue. In preliminary experiments (50) in which I transplanted pieces of salamander skeleton which had been devitalized by quick-freezing and drying, I found that cells of the larval host built a new shell of cartilage onto the dead model (Plate II). Since new cartilage has formed only in direct continuity with the graft, we must conclude that the influence in question is transmitted only by contact and not by diffusion. It seems that objects of this kind would be very favorable for a detailed physical and chemical analysis of communicable differentiation.

In terms of a molecular population concept, this type of inductive effect could be described as follows. The exposed surface of the cartilage contains certain cartilage-specific key compounds. Any potentially chondrogenic cell would contain similar key species in its interior. On making contact with cartilage, these key species would become concentrated and oriented along the surface by virtue of their affinity to the cartilage. After saturating the surface, they would then direct the subsequent chemical transformations in the cell interior into a cartilage-specific course. Thereupon the cell can, in its turn, act on the next one, and so forth. In the case of cartilage, the selective effect would be exerted by the secreted ground substance rather than the cell body itself, but in other instances it would pass directly from cell to cell. This concept has been developed in greater detail in an earlier publication (54). Its main thesis is that adjacent cells influence each other's development by means of selective attraction among specifically configurated molecules, key-lock fashion. An alternative assumption would be an actual transfer of specific master units from cell to cell, in the manner of a spreading infection. Present evidence is insufficient for a final decision.

The second category to be illustrated here is "heteroinduction," in which one tissue induces another tissue, with which it has secondarily come in contact, to differentiate in a direction different from that of the inducing part. Inductions of neural plate by chorda-mesoderm and of lens by retina are familiar examples. According to our hypothesis the mechanism might be as follows. The cells of the inductive layer have a certain specific surface organization, that is, they possess already a surface layer of specific key molecules in relatively stable configuration and orientation. Corresponding molecules of complementary configuration would be present in the ectoderm cells; but during the earlier labile phase of these cells, these would still lie scattered in the interior, mixed with an assortment of other species. As the organized substratum then doubles under the ectoderm, its surface molecules would attract their molecular counterparts in the ectoderm cells into the contact surface. From their oriented surface positions, these molecular layers in the basal face of the ectoderm cells would then control the further chemical developments in their cell bodies, as outlined above. It can readily be seen that the type of response to be obtained from such an inductive action depends both on the character of the substratum and the types of molecular species present in the reacting cells at the time of exposure. The progressive transformation of the molecular equipment of the cell explains why there is often a limited period of "competence" (43) during which cells can appropriately conform to a given inductive stimulus.

In contrast to previous hypotheses, which for the most part have ascribed the inductive effect to diffusible "evocator" substances, the theory here advanced presumes a definite spatial organization of the molecular population as essential for inductive success. Induction of this type could be transmitted only by direct contact. This is in agreement with experience. Lens induction requires intimate adhesion between retina and epidermis. Neural plate arises in direct contact with the chorda plate. Many more examples could be cited that would seem to discount the diffusible nature of the inductive agent. One might object that neural inductions have been obtained in explants floating in body fluids, but it is conceivable that in these cases the active role is played by a protein film adsorbed to existing surfaces to which the explant has temporarily adhered, or a similar film settling along the outer surface of the cells. Since there is a direct relation between the presence of disintegrating cells and the occurrence of neural inductions (18), it is plausible to assume that some neural key species, set free from the destroyed cells, collect and become oriented along the outside of the surviving cells and thereby become an inductive film in the sense discussed under homoeoinduction.

These comments are conjectural. There are not enough facts known on which to decide the issue of contact versus distance action conclusively. It stands to reason that a precise and objective description of the cytological changes attending induction, together with experimental tests of their relevance for the process, could provide some of the wanting clues. A suggestive instance is the following. In observing tissues in inductive interaction I have often been struck by the remarkable correspondence of their cell and nuclear orientation just during the crucial phase. This phenomenon is very marked, for instance, in the formation of the lens. When the eye cup makes contact with the overlying epidermis, the irregular cells of the latter, within an area exactly coextensive with the area of contact, turn into a prevalent radial orientation, each elongate nucleus lining up in the direction of the axis of the opposite retinal cell. Precisely these oriented cells will later constitute the lens placode. The orientation effect is transitory, being obliterated by the subsequent transformations of the placode. While experimental tests remain to be applied, the cell-sharp coincidence between area of contact, orientation, and lens determination makes it almost certain that the observed cell changes are directly related to the primary inductive effect. Evidently, what happens is that the content of the epidermal cells undergoes a thorough reshuffling, with the underlying retinal cells acting as attractive foci. In short, the elongation and orientation of cell structure would be visible expressions of the oriented shifts of the molecular population, which would have to be expected from our theory. A systematic investigation of this orientation phenomenon is under way.

The foregoing comments on induction were presented mainly as illustrations of how problems of development gain in sharpness when resolved into terms of molecular behavior. The special solutions suggested should be considered merely as a first crude and wholly tentative effort to replace the vague, noncommittal, and unrealistic notion of "inductor substances" by a specific, concrete, and testable concept. The limitations of this concept are self-evident. Therefore our insistence lies more on the merits of the particular manner of viewing the phenomena than on the tentative views presented.

As for the limitations, one must not lose sight of the fact that the concept pertains only to the instrumentalities of biological processes and not to the overall order which these processes follow and which we usually refer to as organization. It can explain the manifest differentiation of a given molecular population, provided only that there is a regular frame of organized surface conditions to start with. The organization of the cell content is thus referred back to a prior topographical organization of the various surfaces of cell, nucleus, chromosomes, etc., and the principle, *omnis organisatio ex organisatione* (48), still holds. The "field" character (47) of the organizing conditions in development, which cannot be reduced to molecular terms, has not been touched at all in our discussion.

Another point is that our concept is perhaps too exclusive in assigning the major ordering role to the surfaces. Some faculty of self-sorting of molecular mixtures (as in the formation of tactoids [3]) will perhaps have to be brought into the picture to make it complete. Also our assumption that selective transmission effects and affinities are essentially explicable in terms of steric conformance of specifically shaped molecules remains to be verified. It is still possible that ultimately we may have to take recourse to some other sort of "resonance" mechanism, "resonance-like" interrelations among tissues being clearly recognizable on the biological level (46).

Growth

The problem of growth likewise resolves itself now into a number of tangible issues. The definition of growth gains precision. Growth is often defined as any increase in the mass of an organic system. Since such an increase may occur by the mere intussusception of water—and even a dead seed can swell—this popular definition is scientifically of no value. I propose to define growth in a more restricted sense, as the increase in that part of the molecular population of an organic system which is synthesized within that system. The other part of the population, consisting of inorganic substances like water and salts, passing in and out of the system, is left out of consideration, and so are the smaller organic compounds that enter as ready-made building blocks, e.g. sugars and amino acids, until they have become assimilated and thereby lost their identity. Since, as far as we know, the synthesis of the complex high molecular systems in a cell occurs only in the presence and with the intervention of other similarly complex and cell-generated systems, after their own image, growth in our definition is essentially synonymous with reproduction. In population terminology, we might say that immigration of molecules into the cell space as such does not constitute growth, but their assimilation into complex systems after the model of preexisting indigenous systems does.

We can thus divide the cell content roughly into two categories: systems capable of turning out more of their own kind, and others devoid of that faculty—reproductive and non-reproductive ones. The non-reproductive group includes all the elementary ingredients for the higher compounds, as well as the sterile terminal cell products—membranes, fibers, secretions, pigments, etc. In thinking about these matters it is rather distressing to note that the proteins about which we have the most direct information, such as collagen, keratin, melanin, myosin, pepsin, etc., belong in the latter class and hence can divulge nothing about the reproductive systems, which are still very much of a mystery. At any rate the distinction here proposed enucleates the core of the growth problem, as it were. We shall say of a pigment cell that it grows not when it synthesizes more melanin granules, but only when it synthesizes more of that peculiar part of its protoplasm which, other conditions favorable, can synthesize pigments. Similarly, growth of a myoblast is not the synthesis of more myosin and actin, but of more of the basic muscle protoplasm possessing, among others, the faculty to synthesize myosin and actin; and so forth. For the sake of simplicity let us ignore for the moment the fact that "more" refers not to absolute production, but to the net excess of production over destruction in opposite metabolic phases. It would also be premature to regard the "basic" systems in question as nothing but enzyme mixtures, although enzymatic activity is one of their prime characteristics.

The fundamental feature of growth, therefore, is the capacity of a reproductive system to procreate more systems with similar reproductive faculty. In comparison with this fundamental phenomenon of multiplication of the molecular key population, the often superimposed phenomenon of cell division is as incidental as the setting up of new administrative districts in a growing country. The relation between cell division and growth is by no means as close as the widespread habit of treating them interchangeably would make it appear. During cleavage, an egg undergoes repeated cell divisions with practically no growth at all, while conversely, certain cells, such as mature neurons or the single-celled kidneys of some nematodes, grow to comparatively enormous sizes without division of either cytoplasm or nucleus.

The growth process is being explored from many directions and with increased momentum. Pertinent information comes now from genetics, now from virus research, now from embryology. Views begin to converge, and concepts, to become unified. Yet the more diversified the contributions, the more we must guard against the obvious danger of confusing a common label with a common object. If growth means cell division to one, increase in weight to another, elongation to a third, metabolism to a fourth, protein synthesis to a fifth, their mutual communication on "growth" is apt to become deceptive unless they are fully explicit. I want to make it clear, therefore, that I shall use the term strictly as defined above, that is, as the multiplication of that part of the molecular population capable of further continued reproduction, irrespective of whether or not accompanied by cell division.

With that definition in mind we can now ask a number of specific questions pertaining to the mode, the rate, and the site of growth. First, do cells ever stop growing? Superficially, cells fall into two categories, those that keep on proliferating, and those that do not. The ratio of the former declines steadily during ontogeny. Yet even in the mature body proliferation is still very active in many tissues, such as skin, blood-forming organs, many glands, sex cords, and others. Other cell groups become stationary. Since incremental growth can be observed only if there is a net surplus of production over breakdown, it is quite conceivable that the stationary cells are likewise engaged in growth, but at a rate just sufficient to balance the metabolic losses. In this case, protoplasm would be continually regrown in all cells.

What evidence is there for this view? Our habit of lumping all synthetic processes in one class and all degradation processes in another tends to make incremental growth appear simply as a shift in the balance toward the former; in other words, simply as more of the same kind of activity that is going on in any metabolizing cell. But if we distinguish, as we must, between different levels and types of synthesis, the problem loses its aspect of simplicity. To crystallize the issue let us focus on the most complex molecular systems, the proteins. We know from isotope tracer work (35) that proteins are in a state of continuous renovation. Now, this may mean either that the individual molecule merely interchanges constituent parts with its environment while preserving its identity and individuality, or that the molecule breaks down completely while another whole one, freshly delivered from the reproductive system, appears in its stead. The difference is as between keeping an automobile in repair by replacing the worn-out parts one by one, or scrapping it and replacing it by a new one. If the latter interpretation is correct, protein replacement and reproduction, i.e. true growth, would definitely be two different processes. To the best of my knowledge, the matter has not been finally settled.

Let us then confine ourselves to the cells which are known to grow perpetually (basal skin, lymph, intestine, etc.). Each of these cell types is known to produce each more of its own kind. What does this imply at the molecular level? Again we are faced with alternatives that we can only present, without deciding between them. The question is whether the basic systems of reproduction are the same for all cell strains or whether they change in character with progressive differentiation. Perhaps all cell strains retain a certain germinal core population which remains exempted from differentiation and constitutes the sole source of new protoplasmic compounds; in that case the synthesis of each specific compound in skin, blood, gut, etc. would have to recapitulate the whole course of the ontogenetic differentiation of that cell strain. The new units would be turned out in a primordial mold common to all cells and then gradually pass through a series of transformations after the pattern of the already present skin-, blood-, gut-specific units, as the case may be. The other possibility is that differentiation gradually changes the reproductive apparatus of the various cell strains itself so that the skin-, blood-, gut-specific systems would each be enabled to reproduce themselves directly. If we were to postulate that the genes represent the only self-reproducing systems, and assume, on the basis of genetic evidence, that genes do not change in character during ontogeny (60), the former alternative would become axiomatic. Some new experimental contributions bearing on this problem will be reported later.

Growth Rate

The fact that we do not know just what fraction of the molecular population of a cell constitutes the actively reproductive growth apparatus makes it impossible to determine "true" growth rates, i.e. the rate at which the reproductive fraction procreates more of its kind[7]. This places serious limitations on our interpretation of relative growth, for in all our measurements of growth the reproductive and non-reproductive fractions of the cell are lumped. As long as their ratio remains constant the error in estimating true growth from measurements of total growth is constant and can be discounted for most purposes. For example, if identical cells, kept in different conditions (temperature, nutrients, etc.), grow at different rates, the difference can be properly ascribed to effects on the true growth process. However, if cells of different character grow unequally, even under identical external conditions, this does not necessarily reflect differences of their true growth rates; it may merely be an expression of the fact that different fractions of their total mass are in reproductive activity.

If we designate the reproductive fraction of the population as R and the nonreproductive, hence, constant fraction as D, the true growth rate would be $V_r = \frac{dR}{Rdt}$ while the measured growth rate would be

[7] True growth rate could theoretically be expressed by the time required for a unit quantity of the reproductive system to increase to two units.

$V_m = \frac{d(R+D)}{(R+D)dt} = \frac{R}{R+D} \cdot V_r$. Since $\frac{R}{R+D} < 1$, the measured growth rate will be smaller than the true growth rate by an amount varying inversely with the unknown constant D. Consequently, of two cells with different proportions of R and D but identical fundamental growth rates of the portions R, the one with the larger D will appear to grow more slowly. The fraction D is itself heterogeneous, including water, inorganic compounds, and all organic compounds that have either not yet been incorporated in the reproductive systems or have become detached from them as metabolites and differentiation products. Accordingly the ratio between D and R will be significantly different in different cell types, and hence different actual growth rates must be expected simply as a corollary of differentiation. Whether there is also a concurrent change in basic growth rate remains undecided.

Moreover, the ratio R/D does not remain constant for a given cell strain over its whole ontogenetic history. As differentiation proceeds, the proportion of "differentiated" terminal products, making up a large fraction of D, increases progressively at the expense of R. This increasing drain on the reproductive fraction of the cell is reflected in the well-known decline of measured overall growth rate with age. After our earlier discussion of differentiation it seems hardly necessary to reiterate that the D-fraction in question includes not merely the formed cell products (pigments, secretions, fibrils, etc.) but also that part of their formative apparatus that is not self-reproductive. Even after the loss of the formed products, as in tissue culture, cells of different strains of differentiation exhibit different growth rates (31).

Comparative evaluation of differential growth rates among different cell types is further complicated by the fact that in some cells the formed differentiation products stay within the cell and are therefore included in all measurements, while in other cell types they are extruded and lost. For instance myofibrils, neurofibrils, most pigment granules, membranes, and the like stay with the protoplasms from which they arose, while formed secretions, intercellular ground substances, and the like do not. Comparisons based on visible increments are therefore quite deceptive.

Not until most of these matters have received much more precise formulation and investigation will it be possible to develop a rational theory of differential growth. Attempts have been made to account for differential growth rates by different partition coefficients in the appropriation of nutrients, by competition for "growth substrate," by differences in the efficiency of utilization of foodstuffs, and by differential growth stimulators and growth inhibitors.[8] Plainly, many of these theories miss the point in that they try to explain differences which only appear to be differences of basic growth but in reality are natural results of differentiation. Further compelling evidence on this point will be presented below, where we shall deal with tissue growth.

The Site of Growth

The distinction between a growing (*R*-) and non-growing (*D*-) fraction in protoplasm raises the question of the sites of growth. Evidently, the *R*- and *D*-fractions might be either segregated or interspersed. They might be mingled as mosaics of molecular, submicroscopic, or microscopic (particulate) dimensions; they might be uniformly dispersed or each concentrated in certain localities, which might or might not correspond to the nucleus, the chromosomes, nucleolus, cell surface, or other topographical features.

Some unexpected light has been shed on these problems from recent observations on nerve growth, which I shall briefly relate. A nerve fiber is the long cytoplasmic extension of a nucleated cell body, surrounded by sheath cells. The fiber grows enormously in length and width during ontogeny, and even more dramatically during regeneration. When a fiber is cut, the distal stump, severed from its nucleated base, perishes, while the proximal stump, still connected with its central cell, grows out anew. The rapid elongation of the regenerating fiber is a process of protoplasmic movement rather than of growth in the proper sense (51). But true growth also supervenes and becomes most conspicuous in the increase in width of the fiber (from less than 1 micron to often more than 10 micra, i.e. more than a hundred times in mass). Since the fiber is surrounded by accessory sheath cells and well supplied with blood, one could well imagine that its growth is sustained from local sources. The experimental evidence, however, is that growth takes place exclusively in the nucleated central part of the cell and that all the specific cytoplasm for the whole axon is synthesized there. The evidence comes from constriction experiments as follows (55).

A local constriction (by a ring of artery) reduces the diameters of the neurilemmal tubes which contain the nerve fiber. Fibers which have regenerated through such "bottlenecks" remain permanently undersized from that point on distally. In contrast, these same fibers swell to excessive dimensions just proximal to the constriction. Plate III shows

[8] For some recent examples, consult Needham (29) and Spiegelmann (38).

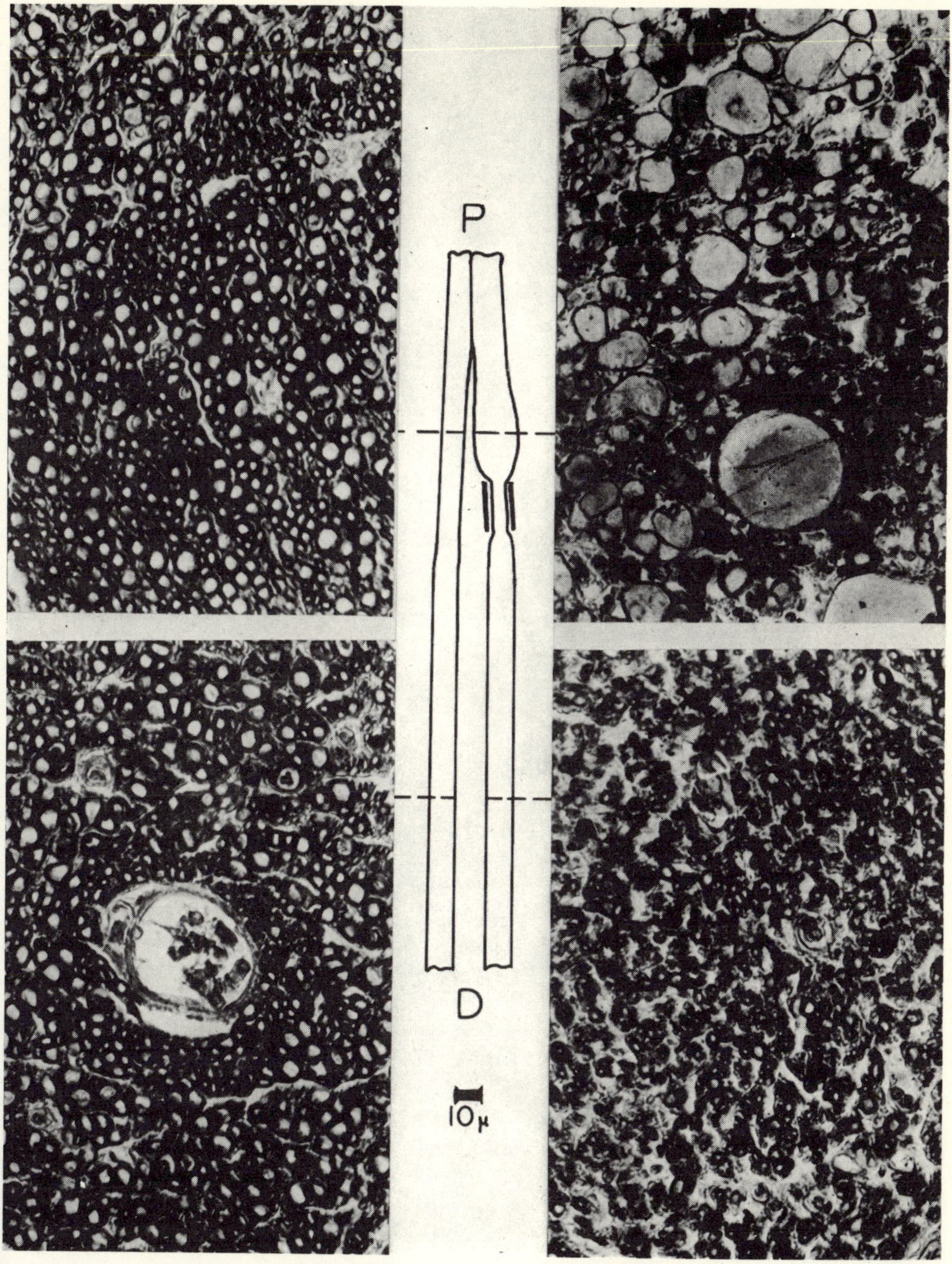

PLATE III. Unequal growth of constricted nerve fibers. Cross sections through a constricted nerve (right) and neighboring control nerve (left) at levels indicated in center diagram, all at identical magnifications (× 425). (Rat, 35 weeks after operation; P, proximal; D, distal). The same nerve fibers which are greatly shrunken distal to the constriction (lower right) have become blown up to giant dimensions just proximal to the "bottleneck" (top right); some fibers have become as wide as a small artery (compare the artery in the center of the distal control nerve at the lower left).

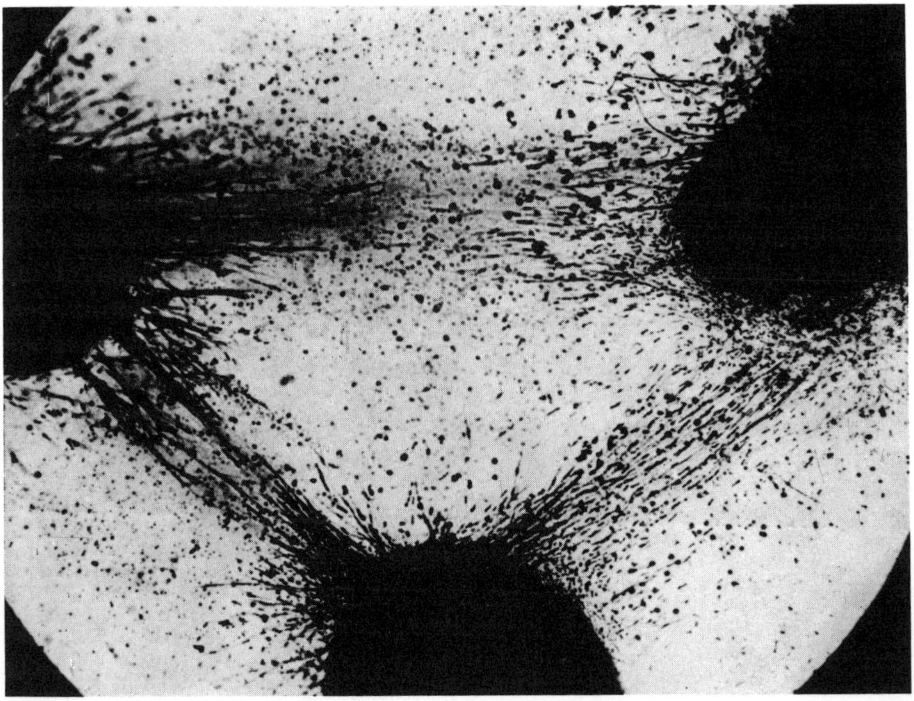

PLATE IV. Oriented cell growth between three explants cultivated in a thin film of coagulated blood plasma. Silver impregnation makes the condensations of oriented fibrin between the centers (compare Figure 3) recognizable as darker streaks in the medium; the cells can be seen to follow these dark bands.

cross sections through an experimental and adjacent control nerve at the indicated levels, located one centrally and the other distally to the bottleneck. A detailed analysis of this phenomenon in a large series of varied experiments has led consistently to the same conclusion, namely, that new axoplasm is constantly produced in the nucleated part of the cell body and that the whole column of axoplasm is in steady slow motion in proximo-distal direction. If this translatory movement is partially throttled by constriction, the supply to more peripheral levels is correspondingly reduced, while the surplus piles up at the entrance to the narrows. If the constriction is later removed, the effect is much as that of opening flood gates; that is, the dammed up material moves downward, its front advancing, according to preliminary determinations, at the order of one millimeter per day.

There is evidence that this proximo-distal growth of the axoplasm occurs not only during regeneration, when it adds directly to the width of the fiber, but actually goes on continuously in all axons, even the mature ones, which have reached stationary size. The central cell body seems to turn out a steady supply of new axoplasm, most of which is then conveyed distad. This raises the puzzling question as to where the material goes, if it no longer adds to the size of the fiber. I submit that it simply serves to replace the protein systems of the cell as they wear off in a steady natural degradation process, and that the nucleated cell body is, in fact, the sole source of such replacement. We know that any isolated fragment of nerve can exhibit the synthetic phases of respiratory metabolism with the aid of the enzyme systems it contains. But in the light of our experiments, it would seem that the enzyme systems themselves cannot be synthesized anywhere except in the central cell body, whence they are then distributed over the whole neuron.

As supporting evidence for this view we can quote the intensive production of nucleoproteins at the nucleus of the nerve cell (20), the identification of nucleoproteins in the axis cylinder (2), and finally what might be taken to be the external sign of their peripheral breakdown, namely the liberation of nitrogen in the form of ammonia from peripheral nerve (15). As a matter of fact, a preliminary estimate of the rate of protein breakdown from the known values of ammonia production leads to a figure which is satisfactorily close to the rate of proximo-distal replacement determined from the damming-and-release experiments reported above. While details of this concept remain to be worked out, it leaves no doubt that the site of growth and replacement of the neuron is strictly confined to the nucleated portion. It is probable, though not proven, that the same would then hold for other cell types.

These facts have an immediate bearing on efforts, currently in the foreground of interest, to bring concepts of genetics and growth in line. Our experiments place the site of growth near the nucleus, not necessarily in the nucleus. We mentioned before that if the reproductive sources should prove to be identical in all types of cells (as is assumed for the chromosomal genes), we would have to assume that their products are instantly recast into specific patterns corresponding to the surrounding differentiated cytoplasm. In our present case the nucleus may continuously reproduce basic protoplasmic systems of the general character dictated by the genic constitution, which, however, under the molding influence of the existing specific neural population to which they are added, would assume specific nerve-character. The alternative would be that the intranuclear extrachromosomal systems themselves have become specifically transformed in the course of differentiation (see p. 144), and in that case the proliferated new compounds (nucleoproteins, etc.) could acquire their full nerve-specific character at their intranuclear site of origin *in statu nascendi.* Our data are noncommittal on this point. They do prove, however, that growth is not a ubiquitous process in cytoplasm and should not be treated at par with the type of synthesis commonly met with in other types of metabolism.

Tissue Growth

Analogous considerations, as applied in the foregoing to cell growth, apply to the growth of tissues and organs. Just as the molecular population of the cell consists of a reproductive and a nonreproductive fraction, so the cellular population of a tissue is composed of proliferative and nonproliferating components. In most organs proliferation becomes gradually confined to certain restricted areas, so-called germinal zones, arranged in layers (e.g. central nervous system, skin), cords (e.g. sex cords) or foci (e.g. buds). Of the newly formed cells, some remain in the germinal zone, while the rest move away from it. The former continue to proliferate, while the latter cease to multiply and usually undergo some terminal specialization.

This dichotomy has sometimes been interpreted as signifying an antagonistic relation between proliferation and differentiation.[9] Statistically speaking, it is correct that as differentiation progresses, proliferation declines. But this does not imply that differentiation in the strict

[9] For a brief summary, see Weiss (47), p. 85.

sense and proliferation are mutually exclusive (11). From our earlier discussion, it should be plain that the proliferating sources of epidermis, bone, blood, muscle, glands, etc. are differentiated to the extent that each can reproduce only its own kind and none of the others. Yet they are still actively multiplying. Only the full transformation into terminal stages seems to be held up in those cells that are kept in mitotic agitation. Proliferation does not interfere with differentiation, but it does impede the elaboration of certain manifest products of differentiation. Conversely, the general impairment of proliferative capacity in cells undergoing terminal specialization can be ascribed to the fact that most of their substance, including mitotic prerequisites, is diverted into the building of differentiation products (e.g. intracellular fiber systems).

In early developmental stages practically all cells proliferate. With progressive differentiation, however, the outlined segregation into reproductive and sterile groups takes place. Since the latter do not further contribute to the increase in mass, the total growth of a tissue or organ will vary directly with the ratio of reproductive to nonreproductive cells. This ratio in turn depends on the geometric configuration of the proliferating zone and the rate at which newly proliferated cells desert it. Further variability is introduced by the fact that most organs receive secondary additions from foreign sources (e.g. blood vessels, nerves, pigment cells), and, conversely, lose some of their own cells by dispersion or destruction. Since each organ has its own rule, it is evident that bulk measurements, which fail to separate the actual sources of the measured materials from the "dead weight," can furnish no valid basis for comparing "differential" growth of parts differing in constitution. If percentage increase is faster in one organ than in another, this could be due to either (*a*) a larger initial proportion of reproductive cells, or (*b*) a larger quota of retention of new cells in the germinal zone, or (*c*) greater infiltration of foreign elements, or (*d*) less dissipation of cells and cell products, or finally (*e*) an intrinsically higher rate of reproduction in the respective germinal cells. Only differences not accounted for by (*a*), (*b*), (*c*) or (*d*) are to be ascribed to (*e*), that is, to a higher "growth rate."

For an illustration let us choose the growing vertebrate eye. Due to its simple form and accessibility it has been a favorite object for comparative studies on growth, with the increase in dimensions serving as index. How heterogeneous a collection of events this index covers can be readily seen from the following account. The original eye vesicle consists of a certain initial allotment of cells from the embryonic brain wall. At first, all of these cells divide. The growth function at this stage is therefore a volume function. In the cup stage the retina becomes multilayered, with a sharp division into a germinal and a sterile zone. Only the cell layer in contact with the outer surface, corresponding to the ventricular (ependymal) layer of the brain, continues to proliferate, while the cells released into deeper layers differentiate the various retinal strata without further multiplication. The source of growth thus has become reduced to a two-dimensional one, causing a marked decline in the relative growth rate taken over the whole organ (e.g. from measurements of diameter). Later, the cells of the germinal layer themselves cease to proliferate and transform into sensory cells, a process which starts from the center (macula) and spreads rapidly toward the periphery (ciliary zone) of the retina. Eventually, only the cells at the rim retain residual capacity to multiply. Further growth is then essentially by apposition from this rim; that is, the growth source has shrunk from planar to linear extension. Meanwhile some of the neuroblasts, though no longer multiplying, grow in size as they sprout nerve processes, which, grouped into plexiform layers, add to the thickness of the retina. During the later stages a gelatinous secretion, supposed to come from cells of both retina and lens, fills the interior with vitreous humor, thereby progressively distending the eyeball. In addition, blood vessels and other mesenchym penetrate into the eye from the surroundings.

This diversity and complexity of the component processes contributing to eye size makes the search for a single "growth-controlling" principle appear utterly unrealistic, and comparisons between different eyes on the sole basis of size must be fallacious. To exemplify briefly the direction that an analytical approach to "differential growth" would have to take, I shall select a few of the concrete questions raised by this picture of eye growth and point out their general significance.

First, what causes the confinement of proliferation to the single cell layer at the convexity of the retina? The answer is unknown. But a clue lies in the fact that in this and most similar instances the site of proliferation corresponds to some prominent geometrical feature of the system, such as a surface or fold or tip or crest. For the cells concerned this means unique physical conditions (e.g. tensions, pressures, unique exposure to surrounding agencies) not shared by the rest of the tissue. In neural tube and retina, for instance, the direct exposure to the fluid of the central canal and ventricular spaces might be the crucial factor. The fact that injury can activate cells of the interior to take part in proliferation[10] could be explained by the penetration of the central fluid to the deeper layers as a result of the break in the internal limiting membrane. This is purely conjectural, but capable of an experimental test.

Next, what determines the further fate of the cells proliferated in the germinal layer? Of the two daughter cells of a germinal cell, either (*a*) both stay in the germinal layer, or (*b*) both move away, or (*c*) one stays and the other moves off. The same alternatives recur at subsequent divisions. Continuation of procedure (*a*) would lead to exponential increase of the proliferative source with no cells left for terminal differentiation, while procedure (*b*) would rapidly deplete the proliferative source. The actual growth of the eye at this phase is therefore determined by the relative incidence of events (*a*), (*b*), and (*c*). But again, factual information on this point is practically wholly lacking. In anuran amphibians the number of dividing cells in the retina has been reported to remain remarkably constant (16), which would mean that all divisions follow scheme (*c*). But this is definitely not true of the chick, for instance, in which the number of proliferating cells increases with the expansion of the retina.[11] Here the rate of growth is therefore largely dependent on the statistical apportionment of the newly formed cells according to (*a*), (*b*), or (*c*). This in turn seems to resolve itself into a matter of the relative orientation assumed by the daughter cells, either during mitosis or immediately after, tangential orientation leading to (*a*), radial orientation to (*b*) or (*c*). A plausible hypothesis is the following.

It has been demonstrated that dividing cells separate in the direction of maximum cell elongation, which corresponds to the direction of maximum tension. Hertwig's rules of cleavage are an illustration of this principle. Assuming that retinal cells follow the same principle, we may conclude that distribution of type (*a*) will result whenever the locally prevailing tension is tangential, while types (*b*) and (*c*) will result from radial tension or lateral pressure. According to this concept the proliferating layer can expand only as long as the retina is under tangential tension. The source of such force is found in the distension of the eyeball by the growing lens and the vitreous body. Retina, lens, and vitreous body would thereby form a self-regulating system operating as follows.

Accumulation of vitreous fluid would stretch the retina, producing a

[10] See for instance the compensatory proliferation of cells throughout the remaining half of an embryonic hindbrain after removal of one half (12).

[11] Recently confirmed by unpublished investigations in this laboratory by Mrs. Eleanor Gould.

prevalence of tangential tensions, which, in turn, would cause a large proportion of newly proliferated cells to stay in the germinal surface according to (*a*). As these additional cells are intercalated, the lateral stretch is reduced and may eventually even be reversed into pressure if the cells begin to crowd laterally; in that case, radial movement (*b*) or (*c*) would result. Meanwhile, however, further secretion of vitreous humor has produced further distension so that more new cells can be accommodated in the surface expanse, and the cycle repeats itself. Because of widely varying local conditions, these events will not cause a regular alternation of phases of tangential and radial divisions as in cleavage, but will merely statistically increase the probability of the occurrence of one or the other. There are some experimental results suggesting such a mechanism of growth regulation. When the vitreous humor of an embryonic chick eye is drained, the collapsed retina remains greatly reduced in size, but increases in thickness (56). The loss of a progressively distending core would properly account for the result. It has also been demonstrated that when incongruous eye-lens combinations are produced by heteroplastic transplantation, a mutual adjustment of size occurs (1, 17). This could be ascribed to the fact that an undersized lens, both because of its smallness and because it will permit vitreous body to flow out, will not be able to exercise the distensive function which promotes retinal growth until it has caught up with the size of the eye and sealed the borders. Reciprocally, an oversized lens will produce greater distension, hence, expansion of the germinal layer, resulting in intensified retinal growth. In a similar capacity, distension of vesicles, glands, ducts, etc. by their own or foreign discharges may be a major factor in regulating the extent of their proliferating layers, and consequently of the amount of growth. Perhaps the excessive size of the gut in amphibians fed on a bulky vegetable diet as compared with meat-fed ones (22) is due to the same principle. One could even envisage that the air in the larval lungs of gill-breathing aquatic amphibians is there for a growth-supporting rather than respiratory purpose.

These examples may suffice to illustrate the dependence of overall "growth rate" on the special mode of growth of each organ, particularly the geometric configuration of its germinal zone.

Orientation of Growth

Growth as such, i.e. the mere increase of the specific molecular population of the cell, is a scalar process. It has no intrinsic direction. Its directiveness is given to it by vector components of the physical framework in which it occurs. Yet growth itself plays a decisive role in creating those physical conditions which, in turn, will orient its further course. A few examples may illustrate this principle of reverberation.

In the preceding section reference was made to the effect of tension on the form of growth in epithelial tissues. Even more spectacular is its effect on the architecture of tissues of mesenchymal origin. Connective tissue, tendon, fascia, cartilage, bone, blood vessels, muscles, all respond to mechanical tensions by orientation of the component cells and intercellular fiber systems along the lines of stress. Experimental analysis of this response in tissue culture showed that the observed cell orientation is a secondary product following a primary orientation of the colloidal matrix in which the cells lie embedded (44). The sequence is as follows: (1) Tension produces alignment of aggregates of the anisodiametric molecules of the matrix into chains. (2) Such chains join to build up submicroscopic fibrils, which eventually may merge into microscopic fibers. (3) Cytoplasmic filaments (filopodia) from cells inhabiting the matrix extend along the interfaces between these solid micellar threads and the interstitial liquid. (4) The rest of the cell body follows the advancing filopodia. In this manner the cell becomes greatly elongated and is made to trace the ultrastructural pattern of its surroundings. Nerve processes advance in the same fashion, except that their cell body remains anchored (45).

In the original tissue culture experiments the matrix consisted of blood plasma, with fibrin constituting the orientable ultrastructural reticulum. Later experiments (53) revealed that even in so-called cultures in liquid medium, cell orientation is brought about by a similar reticulum, namely, coagulable fibrous material exuding from the cells themselves, spreading along the surfaces of the liquid and constituting a trellis ("ground mat") on which the cells then advance. Tensional fibrin organization in blood clots has likewise been disclosed as the mechanism orienting cell growth in wound healing and nerve regeneration in the body (57). In embryonic tissues, other fiber proteins can be expected to operate in like manner. All pertinent observations and experiments made thus far confirm the conclusion that tension, by way of its organizing effect on the colloidal cellular milieu, can determine (*a*) orientation of cell movement, (*b*) cell elongation, (*c*) cell shape, (*d*) orientation of mitotic spindles, hence, direction of division, (*e*) rate of movement, (*f*) rate of multiplication, (*g*) deposition of fibrous cell products. Figure 1 illustrates diagrammatically how different degrees of stretch applied to a fibrous colloid translate themselves into the morphology of inhabitant, as well as immigrating, cells. Potent as these effects are, they could not be a major factor in development if only tensions of extraneous origin were involved. But tensional stresses also arise within the developing organism as a result of shifts, growth, and chemical activity of its parts.

Figure 1. Diagram illustrating the effect of graded stretching (in the direction of the arrows) of a reticular matrix on the shape of mesenchymal cells.

Let us consider the two main intrinsic sources of morphogenetic tensions—expansion and contraction. Their development is illustrated in Figure 2. On the left side (A_o, B_o) are sectors of a tissue possessing a continuous reticulated matrix and containing a circumscribed area (dark circle) in which some localized developmental process goes on, as follows. The lower half ($B_o \rightarrow B_e$) represents a case in which the marked area expands relative to its surroundings, either by rapid proliferation or by deposition of large amounts of ground substance or, in the case of hollow organs, by distension. This local expansion creates in the surrounding continuum tensions that are oriented circumferentially, declining in strength with increasing distance from the border. The fibrous matrix assumes a corresponding pattern as depicted, and this in turn imparts itself to the cells. Cell movements and cell divisions will follow tangential courses. It is evident that the familiar concentric architecture of the mesenchymal sheaths, coats, tunics, and membranes around both solid organs and hollow ducts and vesicles finds a natural explanation on this basis. Essentially the same effect is observed if the surrounding tissue contracts against a rigid core, as in the formation of foreign-body capsules.

A wholly different pattern results if a tissue contains a focal area that contracts. In this case, illustrated in the top row of diagrams ($A_o \rightarrow A_c$), the matrix adhering to the shrinking area is subject to radial tensions, and the surrounding tissue will thus be gathered into a star-shaped pattern. Cell movements and divisions will be directed at the area, instead of around it as in the preceding example. Many embryonic invaginations (e.g. blastopore, neural tube, nasal pits) seem to take their origin from such localized contractile centers in the surface coat of the germ (24).

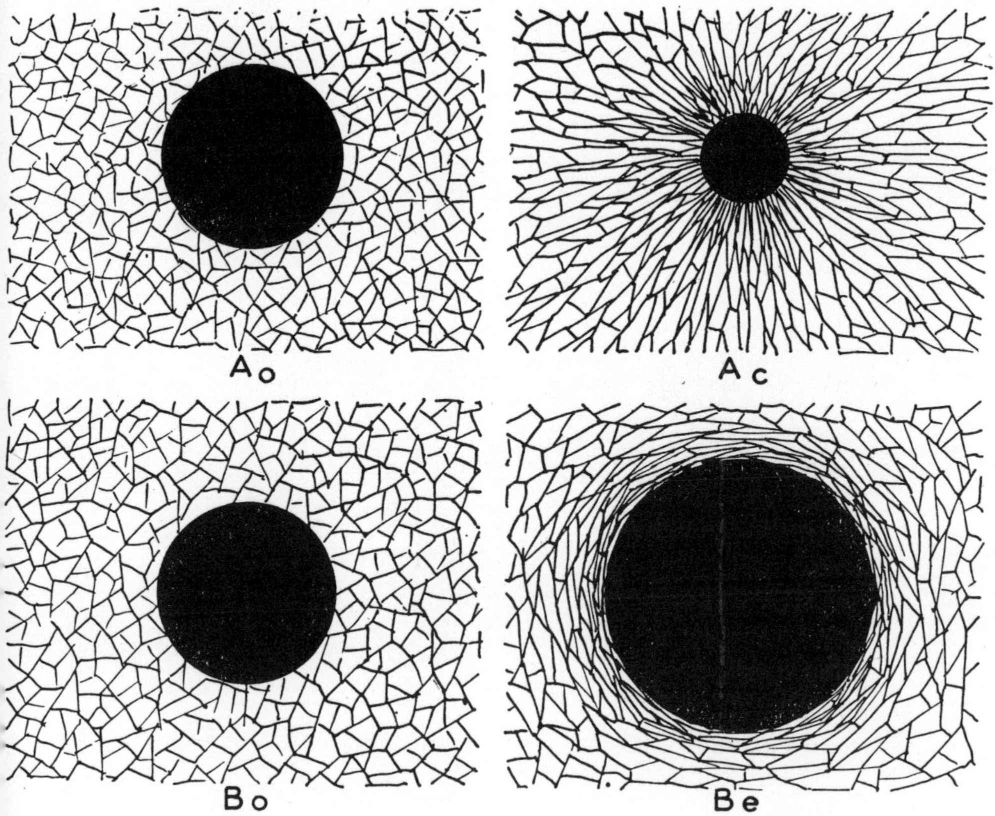

Figure 2. Diagrams illustrating the effect of local contraction (top) and local expansion (bottom) on the structure of a reticular matrix.

It was very illuminating to find that certain biochemical activities attending growth and differentiation may lead to the same type of expansion and contraction patterns (45). The expansion type, for instance, is observed around areas with proteolytic activity. When a local process dissolves part of an erstwhile coherent protein network, the remaining portion, owing to its elastic tension, retracts from the hole and assumes a pattern like that shown in Figure 2 B_e, much like a pierced cobweb. A contraction pattern such as in Figure 2 A_c, on the other hand, arises around actively proliferating centers through the following chain of events. Cells in proliferation, and apparently only then, liberate agents with strong dehydrating potency. The result is a progressive condensation of the surrounding fibrous phase of the medium. Such local condensation in a rigid tissue frame generates tensions, which in turn orient the condensing system toward the center of shrinkage. The familiar ray-shaped growth of ordinary tissue cultures in blood plasma is due to this effect. Analogous configurations can be recognized in many embryonic formations. By its radial orientation such a contraction pattern can evidently guide peripheral cells toward the center. This is one of the major mechanisms by which cells are drawn toward distant destinations, as if "attracted." Its morphogenetic importance becomes particularly impressive when several such centers act on a common matrix, where they create interconnections of great regularity among the otherwise independent centers. As is readily understood from Figure 3, two

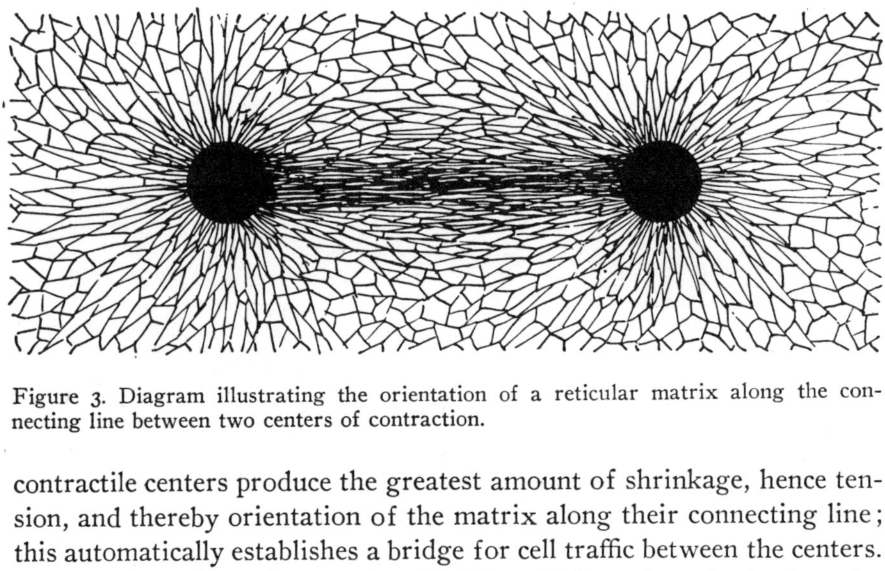

Figure 3. Diagram illustrating the orientation of a reticular matrix along the connecting line between two centers of contraction.

contractile centers produce the greatest amount of shrinkage, hence tension, and thereby orientation of the matrix along their connecting line; this automatically establishes a bridge for cell traffic between the centers. Three centers determine a triangle (Plate IV), and so forth. In principle, these experiments have solved the problem of orientation in cell movement and growth.

In acknowledging mechanical stress as a major morphogenetic factor, two reservations must be borne in mind. Firstly, tension is not the only agent capable of orienting tissue structure. Any other physical force that is capable of affecting the orientation and aggregation of polar molecules (electrostatic fields, electrophoresis, streaming, etc.) may have comparable effects. Secondly, stress is decidedly not a determining factor in cytodifferentiation. Though it seems to be prerequisite for some cell types in order that they may be able to carry out terminal transformations for which they are already predisposed by previous differentiation, it is not the cause of such differentiation. Without stretch, myoblasts cannot transform into muscle fibers (39), but stretch itself neither makes a cell into a myoblast nor ever transforms a non-myoblast into a muscle fiber, as has once been claimed (6).

An interesting combination effect of both contractile tension and proteolysis has recently been observed in the healing of nerve stumps (57). Tension transmitted from stump to stump through an intervening blood clot, intensified by the syneretic contraction of the clot itself, orients the fibrin network in the latter in a prevailingly longitudinal direction. The oriented strands tend to fuse and thus become heavier than their irregularly arranged cross links. Then proteolytic enzymes appear on the scene and rout out the cross links, while leaving the straight linear bundles standing. Thereby preferentially linear order becomes strictly linear.

The main value of these varied results is that they demonstrate in principle how chemical activity (proteolysis, dehydration) can engender spatial order of a high degree, although it has itself no space order. Through its physical and chemical by-products, plain growth thus leads to ordered growth. As the order, arrangement, and consequent exposure of the growing tissue reciprocally affect the condition, manner, and hence, rate of its further growth, we realize how complex and diverse are the factors that contribute to the growth of a given tissue. Some of the factual sequences are summarized in Figure 4. Many more are still unexplored. Carrying this exploration of the "microtechnology" of development, i.e. of "Entwicklungs-mechanik" in its original sense, for-

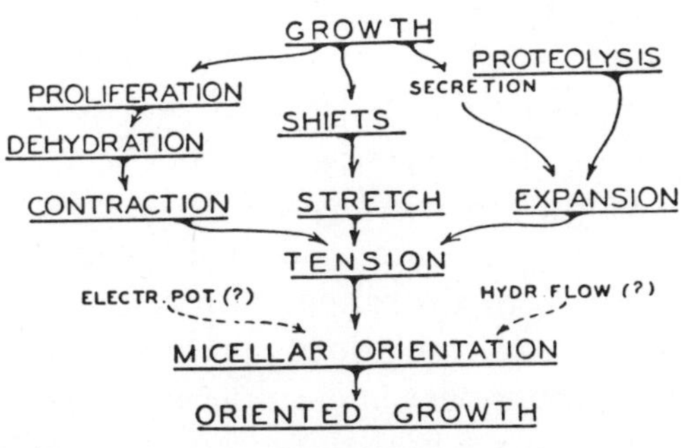

Figure 4.

ward on an intensified scale will go far in rectifying the current rather abstract and often patently unrealistic concepts of "differential growth." But even the existing knowledge, of which we have presented some samples, is sufficiently compelling to discredit the growing fashion of reducing, on paper at least, problems of differential growth to terms of simple chemical reaction rates, with "growth stimulators" and "growth inhibitors" called upon to do the "regulating."

Elaboration of Growth Patterns

According to the preceding sections, differential growth is simply a corollary of differentiation, and differences in growth rates result from a great variety of causes. As such, we have singled out the unequal depletion of reproductive protoplasm in the building of specialized differentiation products in different cell types, the various inequalities arising from shifts, aggregations and dispersals of cell groups in accordance with properties acquired during differentiations, the resultant differences of geometric configuration, and hence, exposure of different parts of a tissue, with the ensuing confinement of proliferative activity to circumscribed zones of varying sizes, the different fates of differentiation products, which are retained in some tissues but extruded in others, and other related disparities contingent upon the divergent courses of differentiation. Such inequalities in the growth pattern can, in turn, secondarily influence the setting for subsequent steps of differentiation, and the resulting changes will further modify the growth pattern. The actual elaboration of the mature body thus appears as a complex sequence of transforming patterns in which the contributions of differentiation and differential growth are intimately interlocked. Basically, however, differential growth is not an agent but merely a product and index of differentiation.

This realization is of the utmost importance when it comes to interpreting the manifold general non-localized influences which modify the growth process during later developmental stages, and which are impressively illustrated by the diversity of phenotypic variants that can be obtained from genetically identical germs: giants, dwarfs, duplications, defects, disproportions, excesses, and suppressions. Since these can often be directly related to aberrations of nutrition, hormones, oxygen supply, physical factors (radiations, pressure), and even climatic changes, all of these agents have at one time or other been claimed as the "dominating" factors in the growth process. We realize now that any factor that has an effect on the physico-chemical setting in which growth occurs may thereby alter the manifest form which the growth process will take. But it is evident that such actions will be in the nature of merely modifying already established patterns rather than instituting wholly new ones. Consequently an agent applied systemically (non-topically) may accentuate or reduce inherent differentials but cannot create differentials not yet in existence. A few illustrations may be in order.

As the developmental realization of the primordial growth patterns of the germ progresses, small initial differences among parts become greatly amplified. Minor deviations from normal become exaggerated. Therefore, if each part were to keep on growing independently of the other parts, the harmony of development would soon be endangered. The setting up of new systemic growth controls coordinating the separate parts thus becomes vital for harmonious development. This is achieved through the humoral, more specially the circulatory, systems for effective and rapid diffusion of substances. But we understand now that a humoral pool could exercise no discriminative control over the various parts supplied by it unless these parts had already previously acquired different response qualities. For instance, the fraction of a common supply that a given part will be able to secure is a function of the accessibility of the part to the circulatory channels; this relation, however, is entirely a matter of previous differentiation and growth of both the part and the vessels. Similarly, the capacity of a part electively to remove, concentrate, utilize, or store certain components from a common pool only reflects affinities of its cells acquired during their prior biochemical specialization; affinity for iodine in the thyroid cell, for vitamin A in the sensory cell of the retina, etc. Or if one cell group responds to a circulating hormone in a manner or degree different from that of another cell group, equally exposed, the two must have been already different constitutionally at the time the hormone appeared on the scene. The case is no different from that of selective drug response of mature cells, which is likewise based on the specific sensitivity of the reacting targets.

Since no nutrient or hormone or other chemical of ubiquitous distribution can make a cell render any contribution for which that cell has not been predisposed by its previous differentiation, none of those agents can be rated as primary causes of differential cell behavior. But they do affect the subsequent elaboration and final proportions of the growth pattern in a variety of ways. A general deficiency of nutrients, oxygen, or specific growth accessories, for instance, hits different growing parts differentially, depending on their respective metabolic rates, margins of tolerance, reserves, substitutive capacities, and the like. Faced with deficiencies, they compete for the available requisites in what Roux has described as the "struggle among parts." One part's loss may either be another part's gain, or may doom other parts, depending on the mode and degree of their mutual dependence. The production of differential growth defects by different vitamin deficiencies (59) may be pointed out as a most instructive example. But the lessons from such deficiencies are not reversible. While increasing the supply of growth essentials from deficient to adequate levels will benefit growth correspondingly, a further increase to levels of overabundance does not keep on enhancing growth indefinitely. A limit will be reached when the production plant works at full capacity, the maximum rate of production being determined by the intrinsic rates of the component processes, which, because of the assembly-line character of growth, are limited by the slowest links in the chains. A fuller discussion of these problems has been ably presented in other chapters of this book.

Of course, since what we call the "normal" level of nutrients, vitamins, hormones, and other growth requisites still falls short of the attainable optimum, the growth potentialities of the organism are "normally" not fully realized. Raising such growth factors to supranormal dosages can therefore occasionally have the spectacular effect of bringing out latent formations which under "normal" circumstances would have remained unrealized. For example, overdosing a female opossum with sex hormones causes the formation of prostate glands (26), a capacity inherent in the urogenital tract but failing to come to expression at "normal" hormone levels.

To the various humoral principles thus far recognized as secondary regulators of growth will perhaps have to be added an even more general category, namely, systems of the antigen-antibody type consisting of organ-specific discharges and their molecular counterparts. The case for this assumption has been presented on an earlier occasion (54), together with supporting experimental evidence. In the most condensed version, it is as follows. The specific key molecules which arise progressively in the course of differentiation (see above, page 153) act as templates ("positives") in the intracellular reproduction of more of their kind. At the same time they cause the formation of molecular species of complementary shape ("negatives"), which may or may not be an integral step in the reproductive process itself (double reversal). Complementary pairs could inactivate each other by conjugation. The amount of free active growth catalysts in a cell would thus vary with the excess of one type over the other. Now, if we assume that the "positives" are retained in the cell (e.g. because of large dimensions), while the "negatives" can escape, circulate, and enter other cells (e.g. because of smaller size), we can see the general outlines of a highly specific and sensitive mechanism of growth regulation, which would coordinate all cells of common character into a single operational system, no matter how much they are scattered. Local reduction (or augmentation) of growth in any part of the system would imply a corresponding decline (or increase) of both "positives" and "negatives" in that locality, and because of their easier diffusibility, of the "negatives" only throughout the system. Since systemic reduction of the "negatives" means an increased ratio of active over conjugated "positives," a compensatory increase of growth throughout the rest of the system will ensue; and conversely, enhancement of growth in one part will automatically entail depression of growth in all other members of the system. Thus the total growth of a given system would be held under joint control of all its components.

That some such mechanisms exist is becoming increasingly evident from a study of compensatory growth reactions. That these mechanisms operate, at least in part, after the fashion of immune reactions is indicated by experiments in which organ-specific antibodies were shown to have selective effects on the growth of the homologous organs (54). But the precise mode of operation here sketched by way of illustration is sheer supposition. Whatever its nature may eventually turn out to be, one must remember that it constitutes merely another contributing factor to the highly intricate system of dependencies and interactions which control the growth process. It is no more the master key to differential growth than are any of the other factors discussed before.

Conclusion

What our whole discussion adds up to is that there is no single master clue to the problems of differential growth or of growth in general. The measurable growth of different parts is so intimately dependent on their peculiar configurations and cellular differentiations, which themselves differ sharply and qualitatively, that a purely quantitative comparison of different growth processes is a deliberate abstraction. There are instances where such abstraction has led to the recognition of striking regularities capable of mathematical formulation (41) (e.g. allotropic growth (19)). Whether these regularities are due to the fact that the compared systems resemble each other closely enough in their mode of growth to be essentially commensurable, or to the existence of some undefined principle of organismic nature, exercising overall control over the multiplicity of heterogeneous components involved in growth, can only be decided after a detailed analysis of the underlying phenomena. At any rate, a purely formal treatment of growth, as is often attempted through the interpretation of growth curves, is only a valuable guide to and supplement of, but never a substitute for, a precise analysis of the different forms in which growth manifests itself.

There can be no research on growth as such. We can only study growing objects. And different growing objects follow different methods. The growth of a differentiating metazoan introduces aspects not presented by the growth of a bacterial culture, and the growth of a higher plant has still other peculiarities. As we have pointed out repeatedly, each individual tissue and organ has its peculiar mode of growth. To know growth we must therefore first break down each one of its manifestations into its constituent elementary processes and then study these and describe them in objective terms. This is a long way to go, but there is no short cut.

In our present primitive stage of knowledge impatient attempts to formulate a general and universal theory of growth seem to have little chance of success. Those general theories that are sporadically being advanced are usually "general" only in the sense that they are first derived from some rather special segment of the growth problem and then broadly generalized by proclamation or mere implication. Consequently they may be sound and pertinent within the area of their derivation but gratuitous and invalid in their illegitimate extensions. Oversimplification of facts and overgeneralization in theory are the main causes of the existing discordance among the several contending concepts of growth. Unless we desist from those practices the confusion is bound to grow, in spite of all the excellent experimental work being done in various areas.

Growth is not a simple and unitary phenomenon. Growth is a word, a term, a notion, covering a variety of diverse and complex phenomena. It is not even a scientific term with defined and constant meaning, but a popular label that varies with the accidental traditions, predilections, and purposes of the individual or school using it. It has come to connote all and any of these: reproduction, increase in dimensions, linear increase, gain in weight, gain in organic mass, cell multiplication, mitosis, cell migration, protein synthesis, and perhaps more. It is gravely inconsistent to apply the most exacting standards of precision to our research data and then proceed to mix into their description and interpretation such vague terminology as this. The mixture can be no more precise than its vaguest ingredient. To deal with growth as an entity, which can be "activated," "stimulated," "retarded," or "suppressed," is only part science, and for the other part, fiction. The less we let our work and thoughts be misled by the delusion that "growth" is basically but a simple elementary process, like a "bimolecular reaction," the faster will be our progress toward true insight into the real mechanisms of development. To promote this more factual approach to the problem of growth, I have tried in this sketchy survey to portray the problem in its natural complexity. If the discussion has helped to make clear what the problems are, it will have served its purpose, even if some of the special interpretations and hypotheses presented may not stand the test of time.

REFERENCES

1. Ballard, William W. Mutual size regulation between eyeball and lens in Amblystoma, studied by means of heteroplastic transplantation. *J. Exp. Zool., 81*, 261, 1939.
2. Bear, Richard S., Francis O. Schmitt, and John Z. Young. Investigations on the protein constituents of nerve axoplasm. *Proc. Roy. Soc. London*, Ser. B, *123*, 520, 1937.
3. Bernal, J. D. Structural units in cellular physiology, in *The Cell and Protoplasm*. Publ. Am. Assoc. Adv. Sci. No. 14, 199, 1940.
4. Bloom, William. Cellular differentiation and tissue culture. *Physiol. Rev., 17*, 589, 1937.
5. Brachet, Jean. Embryologie Chimique. Masson & Cie., Paris, 1944.
6. Carey, Eben J. Direct observations on the transformation of the mesenchyme in the thigh of the pig embryo (Sus scrofa), with especial reference to the genesis of the thigh muscles, of the knee- and hip-joints, and of the primary bone of the femur. *J. Morph., 37*, 1, 1922.
7. Chèvremont, M., and S. Chèvremont-Comhaire. Recherches sur le déterminisme de la transformation histiocytaire. *Acta Anat., 1*, 95, 1945.
8. Doljanski, L. Sur le rapport entre la prolifération et l'activité pigmentogène dans les cultures d'épithélium de l'iris. *Compt. Rend. Soc. Biol., 105*, 343, 1930.
9. Dalcq, Albert. *L'Oeuf et son Dynamisme Organisateur*. Albin Michel, Paris, 1941.
10. Danielli, J. F. The cell surface and cell physiology, in Bourne's "Cytology and Cell Physiology." Clarendon Press, Oxford, 1942.
11. Dawson, Alden B. Cell division in relation to differentiation. *Growth* (Suppl., Second Symposium), 91, 1940.

12. Detwiler, S. R. Restitution of the medulla following unilateral excision in the embryo. *J. Exp. Zool., 96*, 129, 1944.
13. Ebeling, A. H. A pure strain of thyroid cells and its characteristics. *J. Exp. Med., 41*, 337, 1925.
14. Fischer, Albert. *Tissue Culture.* Levin and Munksgaard, Copenhagen, 1925.
15. Gerard, R. W. Nerve metabolism. *Physiol. Rev., 12*, 469, 1932.
16. Glücksmann, A. Development and differentiation of the tadpole eye. *Brit. J. Ophthal., 24*, 153, 1940.
17. Harrison, R. G. Correlation in the development and growth of the eye studied by means of heteroplastic transplantation. *Arch. f. Entw. -Mech., 120*, 1, 1929.
18. Holtfreter, Johannes. Neural differentiation of ectoderm through exposure to saline solution. *J. Exp. Zool., 95*, 307, 1944.
19. Huxley, Julian S. *Problems of Relative Growth.* Methuen & Co., London, 1932.
20. Hydén, Holger. Protein metabolism in the nerve cell during growth and function. *Acta Physiol. Scand., 6* (Suppl. XVII), 1, 1943.
21. Just, E. E. *The Biology of the Cell Surface.* P. Blakiston's Son & Co., Philadelphia, 1939.
22. Klatt, B. Fütterungsversuche an Tritonen. I. Allgemeines. Wachstumsverhältnisse. Darmlänge. *Arch. f. Entw. -Mech., 107*, 314, 1926.
23. Langmuir, I. Molecular layers. Pilgrim Trust Lecture, *Proc. Roy. Soc. London*, Ser. A, *170*, 1, 1939.
24. Lewis, W. H. The superficial gel layer and its role in development. *Biol. Bull., 87*, 154, 1944.
25. Möllendorff, M. von. Über Histiozytenbildung aus Fibrocytenreinkulturen des erwachsenen Kaninchens nach leichten chronischen Reizungen. *Zeit. Zellforsch., 12*, 559, 1931.
26. Moore, Carl R. Prostate gland induction in the female opossum by hormones, and the capacity of the gland for development. *Am. J. Anat., 76*, 1, 1945.
27. Nageotte, J. *L'Organisation de la Matière dans ses Rapports avec la Vie.* Alcan, Paris, 1922.
28. Needham, Joseph. *Chemical Embryology.* Cambridge University Press, 1931.
29. Needham, Joseph. Chemical heterogony and the ground-plan of animal growth. *Biol. Rev., 9*, 79, 1934.
30. Palay, Sanford L. Neurosecretion. V. The origin of neurosecretory granules from the nuclei of nerve cells in fishes. *J. Comp. Neur., 79*, 247, 1943.
31. Parker, Raymond C. Physiologische Eigenschaften mesenchymaler Zellen in vitro. *Arch. f. Exp. Zellforsch., 8*, 340, 1929.
32. Pauling, Linus. Molecular structure and intermolecular forces, in Landsteiner's *Specificity of Serological Reactions.* Harvard University Press, 1946.

33. Runnström, John, and Ludwik Monné. On some properties of the surface-layers of immature and mature sea-urchin eggs, especially the changes accompanying nuclear and cytoplasmic maturation. *Arkiv für Zoologi, 36*, 1, 1946.

34. Schmitt, Francis O., R. S. Bear, and K. J. Palmer. X-ray diffraction studies on the structure of the nerve myelin sheath. *J. Cell. and Comp. Physiol., 18*, 31, 1941.

35. Schoenheimer, R. *The Dynamic State of Body Constituents.* Harvard University Press, 1942.

36. Scott, G. H. Mineral distribution in some nerve cells and fibres. *Proc. Soc. Exp. Biol. Med., 44*, 397, 1940.

37. Spemann, Hans. *Embryonic Development and Induction.* Yale University Press, New Haven, 1938.

38. Spiegelmann, S. Physiological competition as a regulatory mechanism in morphogenesis. *Quart. Rev. Biol., 20*, 121, 1945.

39. Stillwell, E. Frances. Cytological study of chick heart muscle in tissue cultures. *Arch. f. Exp. Zellforsch., 21*, 446, 1938.

40. Thomas, J. André. Des épithéliums en culture et dans l'organisme. *Arch. f. Exp. Zellforsch., 22*, 15, 1939.

41. Thompson, D'Arcy W. *On Growth and Form.* Cambridge University Press, 1942.

42. Törö, E. Über Einpflanzung von Gewebekulturen. *Arch. f. Zellforsch., 15*, 312, 1934.

43. Waddington, C. H. *Organisers and Genes.* Cambridge University Press, 1940.

44. Weiss, Paul. Functional adaptation and the role of ground substances in development. *Am. Nat., 67*, 322, 1933.

45. Weiss, Paul. In vitro experiments on the factors determining the course of the outgrowing nerve fiber. *J. Exp. Zool., 68*, 393, 1934.

46. Weiss, Paul. Selectivity controlling the central-peripheral relations in the nervous system. *Biol. Rev., 11*, 494, 1936.

47. Weiss, Paul. *Principles of Development.* Henry Holt & Co., New York, 1939.

48. Weiss, Paul. The problem of cell individuality in development. *Am. Nat., 74*, 34, 1940.

49. Weiss, Paul. In vitro transformation of spindle cells of neural origin into macrophages. *Anat. Rec., 88*, 205, 1944.

50. Weiss, Paul. The morphogenetic properties of frozen-dried tissues. *Anat. Rec., 88* (Suppl.), 48, 1944.

51. Weiss, Paul. The technology of nerve regeneration: A review. Sutureless tubulation and related methods of nerve repair. *J. Neurosurg., 1*, 400, 1944.

52. Weiss, Paul. Evidence of perpetual proximo-distal growth of nerve fibers. *Biol. Bull., 87*, 160, 1944.

53. Weiss, Paul. Experiments on cell and axon orientation in vitro: the role of colloidal exudates in tissue organization. *J. Exp. Zool., 100*, 353, 1945.

54. Weiss, Paul. The problem of specificity in growth and development. *Yale J. Biol. Med., 19*, 235, 1947.
55. Weiss, Paul, and Helen B. Hiscoe. Experiments on the mechanism of nerve growth. *J. Exp. Zool., 107*, 315, 1948.
56. Weiss, Paul, and R. Amprino. The effect of mechanical stress on the differentiation of scleral cartilage in vitro and in the embryo. *Growth, 4*, 245, 1940.
57. Weiss, Paul, and A. Cecil Taylor. Histomechanical analysis of nerve reunion in the rat after tubular splicing. *Arch. Surg., 47*, 419, 1943.
58. Weiss, Paul, and Hsi Wang. Transformation of adult Schwann cells into macrophages. *Proc. Soc. Exp. Biol. Med., 58*, 273, 1945.
59. Wolbach, S. B., and O. A. Bessey. Tissue changes in vitamin deficiencies. *Physiol. Rev., 22*, 233, 1942.
60. Wright, Sewall. Physiological aspects of genetics. *Ann. Rev. Physiol., 7*, 75, 1945.

VIII. PROBLEMS OF ORGANIZATION

BY J. S. NICHOLAS[1]

The mystical haze that often surrounds the word 'organization' has no place in science. . . . Particulate units at any level are not wholly independent of one another. The relations of the particles are part of the system and it is their behavior in relation to one another that constitutes 'organization.' These relations are amenable to experimental investigation and are usually expressed in some kind of configuration. . . . No particle or unit can be clearly understood or its behavior predicted unless its reactions with others are taken into consideration.

—HARRISON, 1940

IN the many years during which descriptive and experimental investigations have been carried out in the effort to elucidate the steps by which the developing egg gives rise to the structures of the adult there has been little change in the formulation of the main problems of organization. The attention of the embryologist is still largely focused upon the questions arising from the investigation of the problems of localization, differentiation, polarity, movements of parts and the interaction of egg components.

His's expression of the dominant problem of his period, the localization of parts in the egg, has developed into the newer aspect of the localization of potencies for the development of these parts, and also the changes that may be brought about by modifying the interaction of such potencies. This may be on any level of organization whether it is in the cellular layer or even at the tissue level. The movements of parts and the mechanism responsible for cell and layer migrations forms an important phase of the present approach to the fundamental problems given above.

The problems of organization remain as expressed in the last century, the expansion and methodological approach to those problems has changed markedly and the concepts of the past have undergone radical change in response to the present experimental approach. Some of the changes in concept concern the ideas of the integrity of the cell as a functional unit, while chemically the molecular level has evolved as a biophysiological unit and physically the atom with its association of electrons has come into being. Embryologists are still engaged with

[1] Osborn Zoological Laboratory, Yale University.

large units whose extrinsic relations are engaging attention, and in the main they are paying little heed to those intrinsic cellular relations which must be of such dominant import in the final extrinsic reactions which are formative of the patterns through which the organizational structural relationships arise.

The unitary influence of cytoplasm and nucleus in the initiation of cell division was a moot question at the beginning of the century. Since in some of the determinant forms the first division coincides with the later axial relations the problem of nuclear or cytoplasmic dominance in the activation of the primary division becomes a rather vital question. It is clearly recognized today that normally there is an interacting significance of each of these parts in which certain components are actively engaged, but that experimentally there are multiple mechanisms for attainment of the result.

The Problem of Nucleo-cytoplasmic Relationships

In the sea urchin Boveri ('07) has shown that so long as the chromosomal complex of the egg contained a normal complement of chromosomal material development proceeded regularly. Double or half of the normal series might be represented, the quantity apparently having little significance so long as the quality of each of the constituent chromosomes was present. Baltzer ('40) and Hadorn ('36) have been able to eliminate the action of the female pronucleus and consider only the reaction of the paternal pronucleus with reference to egg development. Complete development has been attained, as would be expected from some of the preceding results with invertebrate eggs. Conklin ('31) on the basis of much work has decided that the cytoplasm has a very complex and interesting role in the process of development; his results were indicative of the fact that the cytoplasm itself was probably of a greater independent activity than hitherto suspected. Fankhauser ('34) in his study of the polyploidy has likewise shown quantitatively that the influence of the nucleus must be regarded as being comparatively negligible upon the total process of differentiation.

E. B. Harvey ('36 and '46) has made an interesting series of studies upon the Arbacia egg. She has employed centrifugation in separating the uncleaved egg into four parts in each of which the constituents separate regularly. After the process of separation is complete, the fragments are activated parthenogenetically. Subsequent cell cleavage takes place and may proceed to the formation of as many as 500 cells;

all this may occur from a fragment which contained neither maternal nor paternal chromosomes. There is a limit to development, and while there is a preliminary organization, development ceases shortly after the blastula is formed. In the ('46) contribution the clear quarter containing the nucleus was fertilized after centrifugation and regular plutei were obtained. In this series of experiments the action of the cytoplasmic granules was analyzed. She concludes that the matrix or ground substance is the material of importance and that the cytoplasmic granules can be ruled out as requisite to the continued development here attained. The granules are reconstituted after the blastular organization is reached but not from previously localized granular material.

There is one important observation in Harvey's ('46) paper which is not stressed in the interpretation but which may play an important part in the organizational complex. When the clear quarter is fertilized, soon after fertilization the fertilization membrane usually breaks with extrusion of the oil cap. If, however, fertilization is performed, one to three hours later development may be regular. It may be that time is a requisite factor for the molecular reorientation necessary for the adequate formation of the fertilization membrane. If this is caused to form without the necessary cortical reorganization, the deficiency either qualitative or quantitative in the surface molecular layer manifests itself in a degree of fragility which causes the breaking of the layer and the expulsion of the oil cap. The temporal factor permitting of material organization seems here to be of tremendous import.

While the step by step analysis of the parts of Arbacia eggs has proceeded, it has shown clearly that the capacity of the embryo to undergo its primary organization can be completed under rigorous experimental ablation of many of its components. Certain combinations are necessary for a complete development of the pattern but the pattern itself can be broken down into parts which can be separated in time and action into interrelated series, all combining in the interaction necessary for complete embryo formation.

One clear example of the influence of nucleus in development is shown in the inverse symmetry of snails (Boycott and Diver ('23) and Diver, Boycott, and Garstang ('25)). It will be remembered that they studied a group of left-spiraled or sinistral types with which they conducted experiments breeding the dextral type to the sinistral type and securing its results in the offspring which indicated that there was a definite recessive present so that the genetic results appeared not in the first generation of crossing but in the second generation. There is a skip of one generation before the effect manifests itself. They have interpreted this material as showing that the nucleus has a definite effect upon the cytoplasmic relationship, and also that the cytoplasmic forces cause a lag in the appearance of the nuclear constituents which are responsible for the reversal of asymmetry in the second generation. There is here an interesting link of reaction between cytoplasm and nucleus, one which indicates that there is an interrelationship of rather vital significance for the entire field of early organization. Lillie's ('29) theory of the allocation of the various parts of the organism is well known, a differentiation through segregation and differential dichotomy resulting in a definite localization of the different parts in a rather quantitative way. Their qualities are quantitatively disseminated. The analysis of the process of segregation shows that there is more than a simple quality or quantity which has undergone localization in the course of its *movement* to different regions. A tissue which may become part of one definite organization in one region becomes part of an entirely different set of organizations in another. The change in positional setting is an important process in organization.

An interesting and entirely convincing experiment in which genic control is dominant over all other factors is shown in Robertson's ('42) paper. By transplanting the ovary of the homozygous yellow mouse into black agouti hosts a normal environment was provided for development. The lethal factor, however, appeared in exactly the same way as if the embryos were developing in a heterozygous yellow female. The genic mechanism is clearly responsible for the lethal effect in the homozygous yellow. Just what occurs in the blastocyst at the critical stage at which the lethal characters become evident is not yet known.

The Problem of Egg Constituents

Surface changes in the egg are but an indicator of internal changes which are magnified in the peripheral surface movements. The egg is not a simple sphere with an outer limiting surface on which one who runs may read, but it is a nebulous system in which there are multiple reaction spheres delimited by the relations of movement within an inner and an outer layer.

If such could be held, then the primary arrangements within the egg are due to differences in mobile capacity of the various substances within a comparatively homogeneous system. The internal center is a point of relative homogeneity from which forces emanate with a constantly increasing reactivity towards the periphery. This concept is a 3-dimensional translation of Dalcq's ('38) cortico-vegetal gradient and

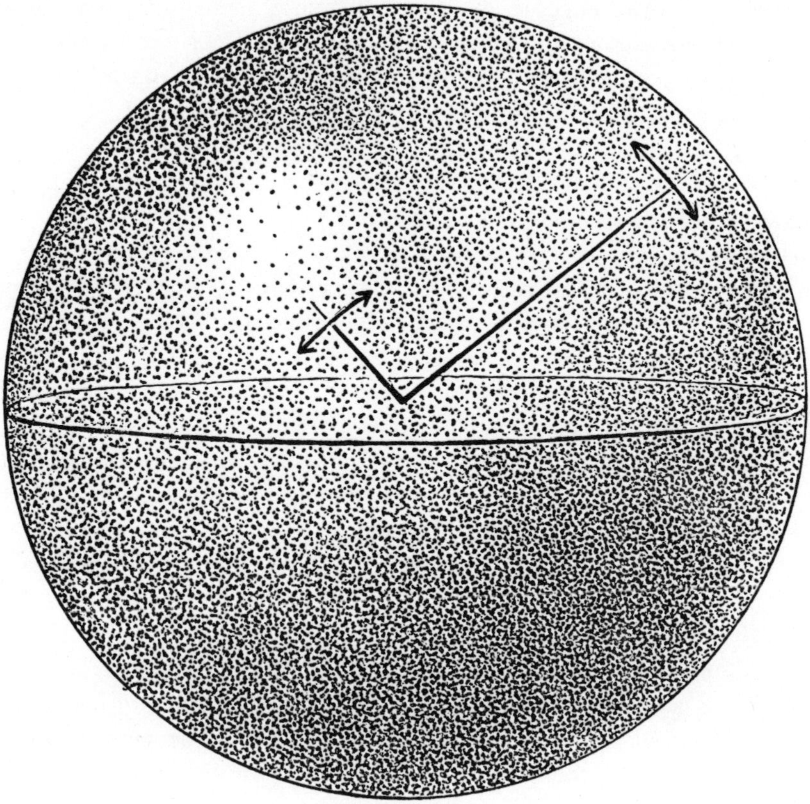

Figure 1.

according to it Dalcq's expression would be simply the resultant of the activities of the multi-sphere relationships expressing their mobile potentiality with reference to each other. Such a concept has the advantage of focusing attention upon the factors incident to the internal configuration of the embryo which have received too little attention.

This thesis does not deny the importance of the surface in the readjustments of polar differentiation subsequent to fertilization. It also would explain the sorting out of processes which occur in the early period as well as the differences in bioelectric potentials which could easily be the resultants of movements of differently charged particles and their reactions during passing phases in which electrical fields might be formed.

Harrison ('45) bases the polarity of the egg upon three factors: the dipole character of protein molecules and their resultant orientation, regional gradational differences in the composition of dipole molecules, and regional differences due to cytoplasmic inclusion. Schrödinger ('45) postulates an aperiodic crystal to account for some of the situations which occur in living systems. Needham ('42) invokes a paracrystalline state or a liquid crystal, but this state would have to be a relatively late one, probably occurring in the early neurula.

The question of gradients within the egg and the effects of different things upon them has long been subject to much speculation. In a recent article Child ('46) discusses the Spemann organizer concept in relation to the various gradient systems. The results are suggestive but not conclusive. The experimental data concerning the Spemann organizer is confined entirely to the amphibian embryo. Child points out that the primary organizer is not primary, since it does not determine external or other axiate patterns in natural development. He bases this idea on Holtfreter's ('44a) and Barth's ('41) work on neuralization without organizer action. He neglects to point out the fact that the amphibiologists themselves early deserted the idea of the Spemann organizer as the only single factor responsible for all organization. While it is exceedingly important in nervous system formation, there are other factors which come into play, some of which have been set in motion far earlier than the Spemann organizer, which cause the movements and reactions of certain parts of the egg. Child has intentionally focused his attention upon the later stages of development in the amphibian and has not mentioned the work which deals with earlier organization patterns.

This is particularly true as we look back and survey some of the literature which has to do with the organization complex in fishes. Morgan in 1893 reported that the yolk might be removed from the egg of Fundulus and that the embryo would still continue to form. Oppenheimer ('36a) has shown that the blastomeres may be removed from Fundulus eggs and that if not removed too early in development the disc will gastrulate and will develop embryonic structures or even complete embryos. Tung, Chang, and Tung ('45) have confirmed these results upon the eggs of goldfish, Carassius auratus. Fragments containing the blastodisc but less than half the yolk formed hyperblastulae. Fragments of the same size derived from 4-cell stages formed embryos. After the 8-cell stage the isolated blastodisc formed embryonic structures or embryos. Oppenheimer ('34, '36b) postulated that some substance or material passed from the periblast during the course of early cleavages. The postulate holds better for Fundulus where complete embryos are not obtained until after the 32-cell stage as compared with Carassius where similar formations are obtained at the 8-cell stage.* Tung, Chang, and Tung consider that the passage of necessary materials is accomplished earlier in Carassius because of the small proportion of yolk present. Lewis and Roosen-Runge ('42) have studied cinematographically the movements in fish eggs (Danio) and find that there is an early movement originating in the vegetal region of the egg after fertilization which up to now has not been considered a general reaction of the differentiating organism. These movements are quite pronounced, moving in a vortex from the vegetal pole towards the center of the egg, resulting in rearrangements of materials at fertilization and in migration of cortical parts definitely through the yolk substance to form a layer, not uniform in its constituency, on top of the yolk between the yolk and the blastoderm. This is a clear indication of the fact that materials in early eggs still have certain visible effects which later are to be correlated with the prospective parts of the developing organism.

The observations of Lewis and Roosen-Runge have their counterpart in the study of rhythmical contractions observed by Yamamoto in a series of papers extending from 1931 to 1938. All of his studies have to do with the early rhythmical movements which are found during the early stages of the egg before segmentation. The significance of these is yet to be interpreted. They may be osmotic in nature but the resultant end point of such a reaction will bring about changes within the egg and cause material dispositions which have a bearing on later organic localizations.

Morphogenetic movements have been plotted in various groups of vertebrates, first in the amphibian by Vogt ('25, '29) in his classic monographs on both the urodele and anuran, then on fishes by Pasteels ('33, '34) and by Oppenheimer ('35, '36b) and later Luther ('36), by Pasteels ('36) for the chick now supplemented by Spratt's ('46) studies with carbon particles, and in mammals by Vanderbroek (unpublished). The vital staining experiments analyze to a large degree the paths followed by cells and by cellular aggregates during the course of their subsequent assumption of position in the embryo.

It gives, however, no index of the effective agents which act upon the

* During the summer of 1948 Dr. T. C. Tung was able to secure complete embryos in Fundulus after isolating the embryonic blastodisc at the 4-cell stage (personal communication).

cells themselves or which cause them to do certain things. It gives a fate map which, while telling the generalized movement of parts, tells nothing about their specific mechanism of differentiation. For this phase of study we are dependent upon other methods, all of which bring about indirectly other effects which sometimes obscure the action of the material itself, or the action of the system immediately surrounding it. This is true of a great many of the transplantation or isolation studies which, while they give interesting pictures of what a cell can do, tell little concerning its reaction in its own specific pattern with regard to its own specific organization. Furthermore, some materials can be mapped less exactly than others. For instance, the teleost blastoderm lacks pigment which is responsible for the maintenance of a persistent color value. Stains likewise fade more rapidly in the teleost than in the amphibian. The same is true in avian studies where the reduction of dye sometimes leads to exceedingly erroneous interpretations. Extirpation or isolation of parts which are to be studied has given quite different results from those secured when the same parts are integral components included within a migrating tissue under normal circumstances and surroundings. It is evident that some change occurs in the early development of the fish egg which is responsible for later developments, since an isolated blastodisc can continue development when isolated after the 8- or 16-cell stage, but before that time it is incapable of adjusting itself to the new environment in such a way as to give a complete embryo formation. Parts may be formed but the total with its organization is not consistently present. Tung, Chang, and Tung ('45) have noted exactly similar conditions in the goldfish egg.

Another factor which seems to have received too little attention is the egg-contained yolk. It constitutes a fundamental problem for each embryo, for the developing structures are molded around the generally spherical yolk content of the egg. The importance of this mass which presents in many forms a surface around which the embryo must grow was pointed out in a paper by Daniel and Yarwood ('39). They studied the distribution of the yolk fragments in various stages of development of the egg using coelomic, ovaducal, and fertilized eggs and compared the stratification of the yolk platelets in each of these stages.

In the sectioned material there is a practical uniformity of distribution at the stages that were surveyed. Studies on later stages show that there is practically no diminution in the size of the yolk platelets until after it is formed in the particular cell in which it is destined to act as a metabolite. Yolk platelets in different systems of the embryo are all of the same magnitude, with the possible exception of the heavy aggregates which are found in the lining of the alimentary tract. This is the last material which is effective in differentiation and is really the beginning of the circulating nutritive factor.

Further study of these stages with the application of the recently developed histochemical tests is necessary for the interpretation of those important agents that are deposited in the various substrates which are present in yolk. Since this is the material which by chemical transference becomes the tissue of the embryo, it is vital that the changes in it be clearly analyzed in order that the relation between chemical changes and the mechanics of development be adequately understood.

Organizational Dependencies

When we come to the study of relationships of organization in separate systems, we begin to perceive how far organization enters into the process and how much is intrinsic to cellular self-determination. Holtfreter's ('39) early studies showed that tissues, when in combination, produce far more than the individual elements by themselves even when acting under an organization complex. When connective tissue is present the cells tend to form more than when endoderm or ectoderm is present by itself. Stableford ('39) found that it was difficult to isolate the internal portion of the system without the necessary adjunct of some of the mesoderm which is going to form connective tissues or muscle layers. He indicates that the endoderm also is a source of mesoderm formation and that isolated endoderm developing in Holtfreter's standard solution can be shown to form mesoderm.*

Nieuwkoop ('46), however, has performed this experiment in reverse and has isolated the ectoderm-mesoderm complex from the endoderm. While there are definite deficiencies in the mesodermal structures, there is an attempt by the organism to produce a complete somatopleure-splanchnopleure combination, but the mesodermal structures are not acted upon by the endoderm which has suffered early removal in the early neurular stage. In these experiments Nieuwkoop has found that two complete sets of limbs will evolve, one of them in the normal location and from the normal structure of a somatopleure, the other formed from the splanchnopleuric portion of the mesoderm at the same level as the somatopleuric limbs. This is the strongest index that we have had

* Stableford in a subsequent study, as yet unpublished, concludes that marginal zone is the sole source of mesoderm and therefore the endodermal capacity for mesodermal origin must be subjected to further test.

so far of what to expect when the action of the endoderm as a molding structure is absent, but when the ectoderm and the nervous system are serving in place of the endoderm. The process of regional localization responsible for limb formation has been able to manifest itself because of the absence of endodermal structures. The endoderm has not been in place for the molding of the internal layer of the splanchnopleure and in consequence the manifestation of limb formation has been expressed in the splanchnopleuric component of the mesoderm. Is this an inhibition brought about by the endoderm, or is it an induction which has moved regionally across in the mesoderm from what was formerly the somatopleure to the splanchnopleure?

In Harrison's ('25) experiments in which the mediolateral axis of the limb is reversed, we undoubtedly have the counterpart to this series of experiments; for Harrison, when he removed the mesoderm underneath the ectoderm comprising the limb bud, undoubtedly removed and reversed the splanchnopleure, leaving the somatopleuric portion internally while the splanchnopleuric was on the outside. From this, perfect limbs developed so that the splanchnopleuric part, already regionally localized when placed in contact with reversed surroundings, cannot effect its normal polarization of limb structures. Whether the somatopleuric layer of the mesoderm has influenced the splanchnopleuric or not cannot be said, but the somatopleure and the splanchnopleure are, in this experimental series, reversible.

That qualities do exist within groups of inducing cells which are comparable to the factors responsible for regional determination in the amphibian has been definitely demonstrated by Eakin ('39) for Salmo. He has divided the invaginated archenteron roof or middle gastrulae of Salmo into an anterior, a middle, and a posterior portion, wrapped each in a tube consisting of extraembryonic half of a late gastrula, and implanted the compound graft of the yolk sac epithelium of an older Salmo embryo. In each case a control graft of the extraembryonic half of a late gastrula was implanted in the same stage as the compounded graft. The control pieces differentiated only into epidermis. When the anterior portion of the archenteron roof was implanted together with indifferent ectoderm, it exhibited little or no power of induction and differentiated for the most part into digestive tube. When the middle portion of the roof was transplanted, it differentiated notochord, somites, gut, and pronephric ducts, inducing the accompanying ectoderm to differentiate neural structures, primarily brain-like in character with auditory vesicles. The posterior portion of the roof, when grafted, also differentiated into chorda, somites, gut and mesonephric ducts; in this case, however, the neural structures induced from the accompanying ectoderm were spinal rather than cranial in character. The three portions of the archenteric roof are thus demonstrated to be essentially similar to the prechordal plate, the head organizer, and the trunk organizer respectively of the amphibian gastrula (cf. E. K. Hall ('37)).

In fish organization before gastrulation the action of the different components is more tentative than fixed, and while the tentative reactions have certain definite sequelae which lead to the formation of organized materials, they themselves are not irrevocably placed in position where they cannot have different attributes from those which normally develop from them. During gastrulation there is a much more specific organization of the blastoderm as a whole. The essential nature of organizing forces operating in the integration of the developmental processes remains to be elucidated by methods not yet at our command. But in teleosts, where the formation of the embryo is limited to a relatively small portion of blastoderm, there are demonstrable integrated activities whose role is the coordination of the various parts of the blastoderm. There is evidence both in Salmo and in Fundulus which suggests the existence of such activity and which indicates that their nature is not identical in the two forms.

The various parts of the blastoderm have different relationships with one another. Luther ('36) divided or sectioned the blastoderm into six sectors and found that not all parts of the blastoderm are equivalent in embryo formation. At all stages investigated the potencies for differentiation as indicated by the percentage of grafts differentiating were highest in the embryonic sector and lowest in the lateral portions of the blastoderm, diminishing gradually around the blastoderm during the course of gastrulation. The differences between the median embryonic and more outlying regions become progressively greater. No differentiation was attained when a central portion of the blastoderm was isolated. Luther interprets the results of these isolation experiments as demonstrating the existence of what he called a physiological field. He postulated that the potency for differentiation in one particular sector was related to the state of physiological activity of its constituent cells, which in turn varies according to the position of the cells with respect to the whole and according to their age. It is highest in the embryonic sector and lowest in the extraembryonic. In the lateral embryonic and in the extraembryonic regions it diminishes progressively during the course of gastrulation. Luther ('37) considered his experiments both by deletion and transplantation methods to strengthen the validity of his interpretation. When a sector of more than 90 degrees was removed from the embryonic area, no embryo formed. In the absence of embryo formation, however, the blastoderm gradually expanded to cover the yolk. When a 70-degree sector was removed from the embryonic area, regulation occurred and embryo formation could still be accomplished by what remained of the blastoderm. When an embryonic sector of 45 degrees was removed from the gastrula stage, the chorda and anterior somites were lacking in some of the embryos which developed. No such gradient has been observed in the egg of Fundulus (cf. Oppenheimer, supra). However, the experiments are not exact counterparts of those on Salmo. Small portions only of the germ ring were removed from regions 90 degrees or 180 degrees away from the embryonic axis and these were grafted into embryonic shields, or into the extraembryonic membranes of an embryo of the same age as the donor. Such grafts implanted on the extraembryonic membrane will differentiate any structures except epidermis, blood cells, and chromatophores. In contrast, grafts from the 90-degree and the 180-degree germ ring of gastrulae of various stages implanted into the shield differentiated without exception provided their cells were incorporated below the epidermis. The nature of the structures differentiated bears no relationship to the sorts of the graft; head, trunk and tail structures were formed in graft from both 90 degree and 180 degree regions of the germ ring. In some cases grafts formed structures characteristic neither of the region of the host to which they were transplanted nor of the regions of the embryo for which the graft itself was originally destined.

Independent Movements

Movements of parts within the egg at early stages are largely determined by the physical reactions of the constituents. During later stages the simpler physical determinants are obscured by specific and generalized processes of growth and differentiation. These processes are chemical and distributional in nature. It has fallen to the descriptive embryologist to relate the surface reactions which are immediately visible, the analysis of the underlying orientations and their role in organization as well as their later definitive action in aligning spatially localized parts or organizers and relating these two components is a fundamental problem of embryology which has so far gone unsolved.

There are many observations and partial attacks upon phases of this problem. Holtfreter ('43b, '44) has recently studied the problem in the amphibian and has shown that many of the movements which we associate with fertilization can occur in the unfertilized egg. The surface movements of pigment and the initial cortical relationships of the plasmogel simulate many of the relations of the fertilized egg reactions. While some of these reactions were observed previously, the various components have not been so completely traced and their significance correlated as in the study referred to. In a second study he has shown the contours of isolated endodermal cells taken from the amphibian gastrula. The blastoporal cells can be identified and spread in the direction of their former proximal-distal axis. They exhibit a definite precocious polarization whereas the non-polarized cells spread in a constant radiating reaction. It must be remembered that these cells are taken from the early gastrula, just as the endoderm is being invaginated. There are two types of cells in the early gastrula, those which are polarized and those which are nonpolarized. Nieuwkoop (see above) assigns to the endoderm the role of generally organizing the mesoderm which later is going to be found in the dorsal mesenteries and in a primordial germ ridge with the general sensitivity for germ cell formation, acting therefore as a primary organizer of this tendency on the part of certain of the mesoderm cells. If this be so, the polarization which Holtfreter has observed has an added meaning, for it is the polarized cells which exert the inductive influence.

In the study of the mechanics of gastrulation Holtfreter ('43a, '44b) has demonstrated the influence of the surface coat which is found upon the cells. Uncoated endoderm is incorporated into an endodermal substrate, whereas a substrate of gill ectoderm will spread over a graft of blastoporal endoderm. Endoderm will pentrate through a hole in the epidermis and pass into an endodermal substrate. This and other factors which are consistent with the formation of endoderm are treated by Holtfreter and material which possesses a surface coat has a definite affinity to invaginate into the substrate. However, when mesoderm is placed upon a piece of glass and two pieces of ectoderm are placed in contact with it, the ectoderm moves up along the outside of the mesoderm and encloses it. In this case the mesoderm is uncoated, while the ectoderm is coated. A reverse effect occurs when a coated piece of the marginal zone is placed in contact with host ectoderm and invaginates under the host ectoderm, which again is a covered type of tissue. In this case we have two covered pieces moving together with the marginal zone to form a type of archenteron. There seems to be no consistency between the way in which the coat acts, but there is a definite specificity in the way in which the tissues act, whether they are coated or uncoated.

Schechtman ('34) discovered the general process of ingression. Schechtman placed stains upon the vegetal region of the egg and found that they disappeared. When he came to examine his eggs further he found that the stain had proceeded upward into the floor of the archenteron going into the blastocoelic floor and that there was a distinct variation as to how far the material would progress according to the stage at which it was placed upon the outside. If the stain was placed upon the unfertilized egg it extended up into the floor of the blastocoele. If it was so placed later, after thirty-two cells, it was just in the lower portion of the floor of the blastocoele, extending to a very slight degree outside the region of the cortex in which it was originally located. The study of ingression seemed so important that it was deemed necessary to carry the study through to the place where Vogt ('29) had stained the early gastrula and followed the movements of particular parts. The study has since been repeated by Nieuwkoop ('46) who confirms, in the main, the results which were brought out by Schechtman, with one minor variation. Since Nieuwkoop stains his materials in many cases considerably later than Schechtman, he got a very distinctly different picture of the final disposition of the stained material. When the stain is applied early, it may proliferate according to extension action effects by which it proceeds directly up the cleavage furrows, the so-called centrifugal effect which Schechtman noticed when he repeated his earlier results. This, however, is not a normal effect, but rather a diffusion one, and the reaction of the egg must be measured in terms of its diffusion constants as well as the changes to diffusion which are undergone according to its degree and state of development during the segmentation stages.

Similar results are attained in my own study of this phenomenon (Nicholas ('45)). In my material, however, there is one step which can be added. There are two distinct steps in the ingression system, the first arising as described by Schechtman and the second occurring in those specimens which are stained deeply and carry a second part of the stain on the surface. It proceeds through the ventral lip and there is invaginated, sometimes by apposition staining the dorsal lip. It is most prominent in the floor of the archenteron and eventually merges with the material which was formerly in the blastocoelic floor forming a continuum with it. The primary ingression is blastocoelic, the secondary a gastrocoelic floor contribution. The first ingression forms the anterior portion of the gastrocoelic floor by an inversion of the blastocoelic floor in the region in which the heart later forms. The gastrocoelic invagination extends forward to the region of later liver formation. These observations indicate mass movements of material which are responsible for regional localization of determination reactions. This is corroborated by Nieuwkoop's results with a slightly different staining method applied to different stages.

Mass Movements in Organization

Two examples of how mass movements are effective in organizational phenomena will suffice. The first is the case of localization in the chick, in which the differences between the older concept of the regional differentiation of the parts of the blastoderm can be contrasted with the results as secured from experimental data.[2]

The His concept of the chick blastoderm was that of a central axial localization in which the regional localizations are arranged in linear order. The Rudnick diagram (even with the necessary modifications imposed by Spratt's work) deserts the axial type of linear arrangement in favor of a distribution of potencies for the formation of the parts of the organism. Thus one finds heart, thyroid, liver, and chorda not spatially localized in linear order but originally as potency areas which later join to express the final types of specific organization.

The second example will be presented in a more complete form, for it is the rather complete story of the localizations anticipatory to organization in Drosophila.[3]

The egg nucleus lies close to the dorsal side of the egg; meiosis leads to formation of three polar nuclei which fuse to form a residual body under the dorsal surface near the female pronucleus. Following fertilization the zygote nucleus divides a number of times and the nuclei are dispersed throughout the deutoplasm of the egg mass. The first migrations or movements within the Drosophila egg are those towards the posterior region of the egg where the pole cells later form. Two to five pole cells are formed, each one of them a direct descendant of one of the nuclei which lay originally within the deutoplasmic mass. These subsequently divide to form a total of about thirty pole cells. After a latent period

[2] Since the above comparison was made, Spratt ('46) has published his results with carbon particles employed to study the movements of parts of the chick blastula. The Rudnick diagram is therefore now subject to modification, but in the main shows the cardinal localization of parts in the chick.

[3] For this information I am deeply indebted to my colleague, Dr. Poulson.

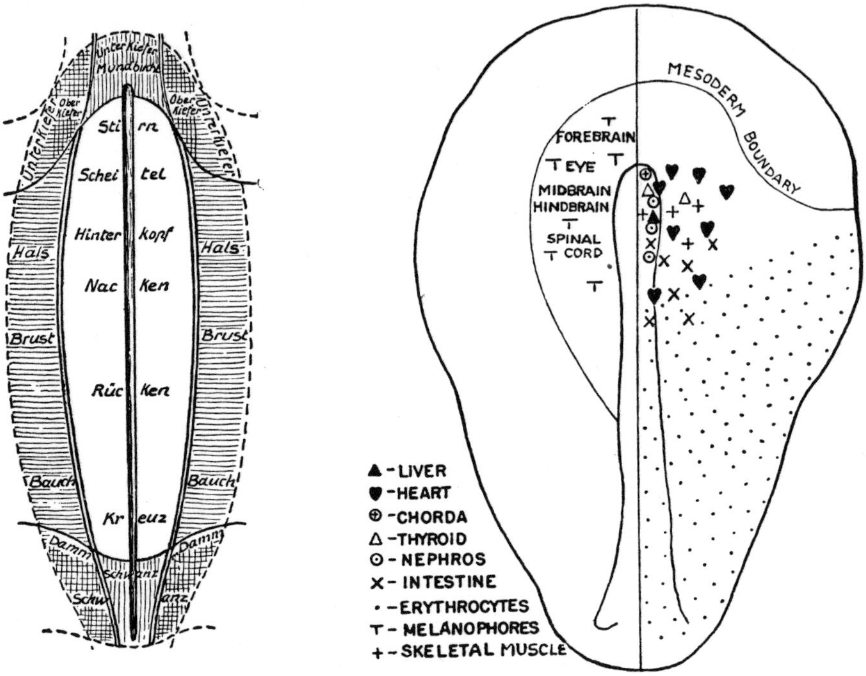

Figure 2. The His concept of localization in the chick blastoderm compared with Rudnick's summary of present findings in this form.

most of the remaining nuclei rather suddenly emerge from the central deutoplasmic mass and are regularly arranged in a peripheral position, where they remain, forming a single boundary layer completely enclosing the remainder of egg materials. The pole cells remain separate from the underlying blastoderm.

The events described can be regarded as the first generalized movement of material within the egg of Drosophila. The deutoplasm still contains a few nuclei, the yolk nuclei, which for some reason did not move towards the periphery. During this process the fused polar bodies have undergone a series of transformations during which the nuclear cycle is synchronized directly with that of the blastoderm nuclei. Eventually the polar body disappears and its chromatin disintegrates and is resorbed into the general deutoplasmic mass.

After the nuclei have taken up their peripheral position and undergone three mitoses, they incorporate some of the egg cytoplasm immediately beneath them together with the cortical cytoplasm into a deep layer which will become cellular material. Shortly afterward, cytoplasmic cleavage begins as an infurrowing from the cortical layer. This separates the blastoderm nuclei and incorporates each plus cortical material and material which came from the underlying stratum of cytoplasm into individual cells. The peripheral cells undergo a growth by accretion or by enlargement of their mass through the incorporation into them of material which formerly lay in the cytoplasm of the central syncytium.

Upon completion of the blastoderm the first movements are those of cells along the ventral mid-line and in the regions which later become the cephalic furrow. They are slipping movements and appear to be consequences of changes in cell shape and polarity. These movements lead to the inturning of the mid-ventral cells to form a mesodermal tube and to the establishment of the proctodaeal plate beneath the pole cells. These movements are neither immediately preceded nor accompanied by mitotic activity. Anteriorly the mesodermal tube extends beyond the cephalic furrow and terminates as a T-like structure just posterior to the region which will produce the stomodaeal plate. At its posterior end the mesodermal tube is for a short time confluent with the depression which gradually deepens in the proctodaeal plate to receive the pole cells. The pole cells subsequently migrate through the proctodaeum and separate into two groups which become the gonads. This period of special movements is followed by one of general movement in which the newly established germ band elongates and extends around the posterior pole of the egg onto the dorsal side, while the dorsal and lateral regions of the blastoderm fold and stretch in making way for the elongating germ band. During this period the proctodaeal plate invaginates deeply and mitoses are found among its cells. Simultaneously at the anterior end certain cells at the level of the stomodaeal plate migrate beneath the ectoderm and undergo mitosis to form a group of cells which become the anterior rudiment of the mid-gut (endoderm). The origin of the posterior rudiment of mid-gut is somewhat obscure, but probably of a similar nature, as there is no evidence of its formation from the mesodermal tube. With the invagination of the stomodaeal plate the anterior rudiment of mid-gut begins to extend by mitosis and cell movement posteriorly and ventrally between the mesoderm (which has now become a flattened layer of cells with no remnant of the original lumen) and the yolk. A similar growth and extensions of the posterior rudiment lead eventually to their fusion to complete the mid-gut.

A period of special cellular movement begins during the invagination of the stomodaeal plate, when neuroblasts begin to be differentiated and move from the ectoderm lateral to the ventral mid-line into the interior against the mesoderm. The neuroblasts contain a large amount of cytoplasm and can be recognized as constituents which are distributed along the entire length of the germ band. The neuroblasts immediately undergo the characteristic unequal divisions which produce the ganglion cells. The entire nervous system is originally bipartite, a condition reflected in its later bilateral structure. The differentiating ganglion cells send out processes from the nerve cord into the mesoderm and these extend along with the mesoderm into the lateral areas, becoming the segmental nerves.

Lateral ventral cells which remain at the surface after the neuroblasts have moved to the interior become hypoderm cells and later produce the larval integument, except in certain localized regions where they give rise to salivary plates (later invaginated and pulled into the interior through the stomodaeum) and the eleven pairs of tracheal pits which give rise to the major part of the tracheal system.

By the middle period of development (10-12 hrs.) the principal larval structures, or their rudiments, are all laid down. The shortening of the germ band brings many cells back to somewhere near their original positions, but there are new shifts as the dorsal closure of the embryo is completed and segmentation proceeds. Special movements are concerned in the involution of the head region, which is characteristic of the larvae of higher Diptera. The imaginal discs begin to make their appearance only a short time before the hatching of the larva from the egg (at 20-24 hrs. after fertilization, depending on the temperature).

The movements within the embryo of Drosophila are more complex than in the amphibia where they are limited to two general types of movement. The first of these is found in the egg just after fertilization, continuing through the segmentation stages and blastulation, the second comprises the later general movements, resulting in the formation and distribution of the layers and the formation of the nervous system. Specific movements of the type performed in Drosophila seldom occur. In Drosophila the specific movements apparently are those of aggregates of cells which tend to adhere to each other and form functional masses from which are derived the specialized tissues and organs of the embryo. An amphibian would produce its nervous system if it were similar to a Drosophila by the creation of more parts from which certain areas would be deleted. Drosophila, however, has its parts fully formed, its ganglion cells sending their processes out to the periphery, and then a contraction or aggregation of the different parts of the system occurs, with a degeneration of some of its parts to form a smaller unit. This method is just as efficient in its complex potentiality as is that of the amphibian.

There are four types of movement that one may designate as occurring in the Drosophila egg. The first are general movements involving nuclei and their peripheral distribution; the second, specific movements of cells and of layers; the third, specific movements of cell aggregates which later will form organs within the organism; and finally the general movements involved in the shortening of the embryo, which continue within the nervous system and result in the high degree of condensation characteristic of that system in the higher Diptera. The condensation of the nervous system continues after the germinal band has begun to proliferate upward along the sides to surround the intestinal canal, and nerve fibers accompany the migrations of the mesoderm. In this it is not unlike the vertebrate system. It differs from it, however, in that (during the process) it sends proportionately very much longer nerve fibers in connection with the contraction of the central system itself. The localization of the various parts and their relationships to future organization are clearly shown in Poulson's ('45) rather remarkable investigation of this form.

The movements described above may be reduced to a rather simplified form in the following classification.

Classification of Movements in Drosophila

A. General movements (early egg)
   1. Nuclear
      a. early distribution
      b. peripheral migration
   2. Cytoplasmic
      a. early (with nuclei)
      b. disengagement from yolk
      c. furrowing of peripheral cytoplasm
      d. syncytial layer to cellular layer

B. Special movements
   1. Cellular migrations
      a. endoderm in gastrulation
      b. pole cells
      c. neuroblasts
   2. Aggregate movements
      a. mesoderm in gastrulation
      b. proctodaeum

c. stomodaeum
d. mid-gut
e. nervous system

C. General embryonic (germ band and later)
1. Surface movements
a. germ band extension
b. germ band shortening
c. dorsal closure
2. Invagination
a. involution of head

The story of the movements of parts during organization is not nearly so complete in the amphibian (Gillette ('44)). Here a similar classification scheme might be evolved, but there are tremendous gaps in the analysis of the various types of movement which occur. There are many observations which indicate that movements occur, but whether these are molecular, particulate, cellular, or aggregate is in general not noted, and accurate measurements of the amount of movement and the forces necessary for the final position are not yet available. Gillette ('44) has investigated the neurula and has given the most concrete example of mass movements of cell aggregates in the formation of the neural plate. There is a dorsal concentration of cells which formerly occupied a lateral position. At the same time there is a smaller but definite relocation of some of the ventral cells which move into the lateral position. There is a loss of 11% of the total ectoderm from the lateral lunes and 7% from the ventral lunes in providing the ectodermal material which is taken over in the dorsal region.

Chemical Organization

The work dealing with the chemistry of the neural organizing substance, or substances, beginning with Spemann, Fischer, and Wehmeier ('33) and continuing to the present (for reviews see Needham ('42) and Boell ('42)) has shown how many different substances may cause this effect. No particular one, however, can be said to be dominant in the reaction. The analyses for glycogen in different regions by Woerdeman ('33), Heatley ('35) and Heatley and Lindahl ('37) showed that glycogen disappears in the dorsal lip during gastrulation. This fact has been employed by Boell ('42) as showing the significance of the reactions in association with the dorsal lip and carbohydrate metabolism. Boell (*loc.cit.*) implies that the quantitative differences in metabolism

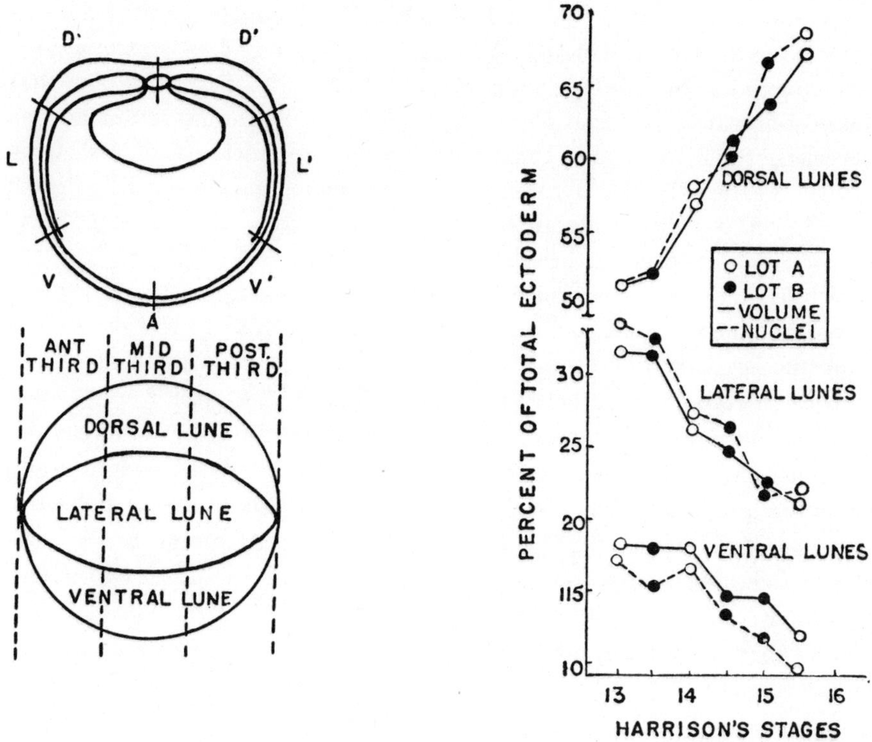

Figure 3. The left hand figure shows the (A) typical cross section of the neurula with the ectodermal subdivisions marked, the (B) is a diagram of the left side showing the subdivisions of (A) reconstructed as lunes on the lateral surface. The right hand figure shows Gillette's graphs of the cellular redistribution of the ectodermal cells during neural plate formation.

in the various parts of the gastrula to be of secondary importance in the inductive action.

Brachet ('38) in cyto- and histochemical studies localized the SH groups in various stages of development. With the exception of the two cell stage, the SH groups are predominantly situated in the animal region of the embryo and during gastrulation are strongly centralized around the dorsal lip, after which there is a heavy localization of these substances in the nervous system. In his 1944 presentation of the various lines of chemical investigation, Brachet shows that the content of ribonucleic acid decreases in the dorsal lip during its invagination (cf. Fig. 94 in Brachet ('44)). There is a coordinate change in the distribution of the ribonucleoproteins and a marked shift in their localization during neurulation. This is coincident with the pattern as shown in his Figure 92.

On the basis of his own work Brachet ('44) concludes that the nucleus is the center of protein synthesis, in contrast to Caspersson ('41), who concludes that the nucleolus is the initiator of the reaction and that the proteins are synthesized in the cytoplasm while the nucleolus acts as the center for histones and the nucleus for the ribonucleic acid. Brachet's studies on the roles of thymonucleic, ribonucleic, and desoxyribonucleic acids in cell metabolism tend to strengthen his viewpoint concerning the character and action of these specific substances.

While there have been numerous studies of oxidation rate in the developing embryo, there has been little correlation of the facts so derived with the organizational picture. Boell's findings indicate that carbohydrate metabolism rather than respiration is the major significant factor. There have, however, been definite correlations between the development of definite chemical entities with both morphological and functional development, and a few examples of this phase of the work will be given. Sawyer ('43) has shown the rate and amount of development of cholinesterase and has correlated this with the onset of the embryonic nerve-muscle reaction.

The enzyme remains constant in quantity during the early stages, deriving its activity in causing nerve-muscle reaction by the concentration of practically the total amount of cholinesterase in the nervous and myotomic structures. The increase in content begins with stage 36, when the behavior pattern demonstrates the S-flexure, after which the enzyme grows in quantity such that it can be represented as a sigmoid curve up to the feeling stage. By placing Amblystoma in solutions of prostigmine and eserine, Sawyer was able to change the rate and amount of the cholinesterase development. In a further study Sawyer ('43) has studied the cholinesterase concentration in the nervous system and in the total animal, showing the increase in concentration during the period of functional organization.

The cholinesterase activity reaches its peak in Amblystoma about fifty days after fertilization in both nervous system and muscle and then declines in both, but more drastically in muscle than in nerve. The adult activity is represented by the same concentration of cholinesterase that is found in some of the pre-feeding stages. When the spinal cord of the embryo is removed so that the muscles develop in denervated condition, the curve of the development of cholinesterase is much lower in the muscle than in the innervated muscle. There is, however, a residuum amounting to 31% of the normal in late swimming stages, showing that the nervous system is not entirely responsible for the organization of this functional unit.

The results of Sawyer ('43) have been checked by Boell and Shen ('44) in their study of the cholinesterase content of induced neural structures.

The results of measuring the cholinesterase content of various tissues in developing the embryos of Amblystoma punctatum indicate a significant localization of cholinesterase in the nervous system in comparison with the ectoderm at the period of development marked by the closing of the neural folds. As differentiation of the nervous system proceeds, the concentration of cholinesterase in neural tissue becomes progressively greater.

The cholinesterase activity of secondary or induced neural structures produced by the action of implanted chorda mesoderm upon competent ectoderm is of the same order of magnitude as that of the primary nervous tissues of the host and is much larger than that of ectoderm which has not received the stimulus of the inductor. It appears then that the phenomena of induction, in addition to causing the development of distinct morphological changes in the induced tissue, stimulates also the latter to develop the biochemical machinery which is characteristic of normal nervous tissue. They suggest that cholinesterase content of secondarily induced neural structures can be regarded as the measure of their potential or incipient functional differentiation.

Boell ('45), in his study of the correlation of respiration and the presence of cytochrome oxidase, finds that the respiratory rate increases steadily throughout development and varies exponentially with time, although it seems impossible to correlate specific morphogenetic effects with abrupt changes in velocity of oxygen uptake. The rate of respiratory increase is not constant and shows a distinct break at stage 30. The evidence supports the conclusion that the rise in respiratory rate noted during development is the result of growth of the embryo, i.e. the transformation of the inert yolk into metabolically active material. The respiration of Amblystoma punctatum embryos is markedly susceptible to cyanide throughout development, but is especially so in the later stages.

As in the case of respiration, the cytochrome activity of Amblystoma punctatum embryos increases steadily throughout the entire period of development and apparently varies directly with increase in metabolically active embryonic mass. Various lines of evidence suggest that respiration is largely mediated by cytochrome oxidase throughout development, but the absolute rate of oxygen uptake at any given stage is limited by some factor other than the concentration of cytochrome oxidase. By the application of the heterauxetic equations the relative growth rates of cytochrome oxidase and cholinesterase have been determined and the individual character of these processes of biochemical differentiation shown.

The evidence in the work of Sawyer, Boell, and Shen shows clearly the place which the development of the enzyme plays in the functional organizational development. The fact that the enzymes have different developmental rates indicates the significance which they have in the general growth properties of the embryo and, what is still more important, how they are correlated with the specific parts of differentiation. The interesting problems of the changes in the substrates from which these materials are formed still await solution and will undoubtedly form an important phase of the understanding of organization.

Regeneration

The problems of organization are not limited to those which can be attacked with embryonic material, for in every regenerating structure there is a new organization involving both growth and differentiation in the attainment of the new structure. It is impossible here to cover the wealth of vertebrate and invertebrate material which yields information on various phases of the problem of morphogenesis during regeneration. One example can be cited in order to show the changing attack in this field, the relation of the nervous system to limb regeneration.

Schotté ('23, '26) on the basis of a great deal of work in the amphibian nerves came to the conclusion that some elements of the nervous system were necessary in order that regeneration should be initiated and completed. This work has been continued (Schotté and Butler ('41), Butler and Schotté ('41), and Schotté and Butler ('44)). They have studied the various phases of the formation of the blastema and the way in which each is affected by the presence of nervous system. The nerves seem to be necessary for the dedifferentiation phase leading to blastema formation and also in the transformation of the blastema into a regenerate with morphogenetic determination. After morphogenetic determination is established, the presence of nerves is no longer necessary for subsequent growth and differentiation.

The research referred to above seems to answer one of the problems of regenerative organization in that it solves the particular role of nervous tissue in blastema organization. That there are other factors still to be analyzed is shown in Rose's ('44) study of anuran forms, where the cuticle forms a block to the regenerative process. By treating with NaCl the stumps of amputated limbs in fully developed frogs or in frog tadpoles after the normal time of regeneration, Rose was able to initiate limb development which was sometimes complete, and at other times with deficiencies in the hand.

Singer ('42, '43, '46a, '46b, '47a, and '47b) has analyzed the amount of nerve tissue necessarily present in order to initiate regeneration. He finds that the process is non-specific and that the number of nerve fibers, whether autonomic, sensory, or motor, controls the reaction. In this way the regenerative process of urodeles is reduced to a quantitative expression of nerve fiber relationship and our concept of the significance of nerves to regeneration is clarified.

In the above pages a few of the problems of organization have been presented. They constitute but a sample of the vast array which have at one time or another been investigated and for which at present there is no final answer. The work here given, however, shows the present attack on problems many of which were formulated long ago. It must be admitted that their final solution has not been attained, but we have reason for optimism, for science works slowly but surely toward the solution of any major problem, such as embryonic organization has proved to be.

REFERENCES

Baltzer, F. Über erbliche letale Entwicklung und Austauschbarkeit artverschiedener Kerne bei Bastarden. *Naturwiss, 28*, 177-187, 1940.

Barth, L. G. Neural differentiation without organizer. *J. Exp. Zool., 87*, 371-383, 1941.

Boell, E. J. Biochemical and physiological analysis of organizer action. *Growth*, Vol. 7 (Suppl.), 37-53, 1942.

Boell, E. J. Functional differentiation in embryonic development. II. Respiration and cytochrome oxidase activity in Amblystoma punctatum. *J. Exp. Zool., 100*, 331-352, 1945.

Boell, E. J., and S. C. Shen. Functional differentiation in embryonic development: I. Cholinesterase activity of induced neural structures in Amblystoma punctatum. *J. Exp. Zool., 97*, 21-41, 1944.

Boveri, Th. Zellenstudien VI. Die Entwicklung dispermer Seeigeleier. Ein Beitrag zur Befruchtungslehre und zur Theorie des Kerns. *Jen. Zeitschr. f. Naturw., 43*, 1-292, 1907.

Boycott, A. E., and C. Diver. On the inheritance of sinistrality in Limnaea peregra. *Proc. Roy. Soc. London, (B) 95*, 207-213, 1923.

Brachet, J. La localisation des protéines sulfhydrilées pendant le développement des Amphibiens. *Bull. Acad. Roy. Belg.*, 1938:499-510, 1938.

Brachet, J. *Embryologie Chimique*. Paris, Masson et Cie., 505 pp., 1944.

Butler, E. G., and O. E. Schotté. Histological alterations in denervated non-regenerating limbs of urodele larvae. *J. Exp. Zool., 88*, 307-341, 1944.

Caspersson, T., H. Landström-Hydén, and L. Aquilonius. Cytoplasma nucleotide in eiweissproduzierenden Drüsenzellen. *Chromosoma, 2*, 111-131, 1941.

Child, C. M. Organizers in development and the organizer concept. *Physiol. Zool., 19*, 89-148, 1946.

Conklin, E. G. The development of centrifuged eggs of ascidians. *J. Exp. Zool., 60*, 1-119, 1931.

Dalcq, A. M. *Form and Causality in Early Development*. Cambridge University Press, 1938.

Daniel, J. F., and E. A. Yarwood. The early embryology of Triturus torosus. *Univ. of Calif. Publ. Zool., 43*, 321-356, 1939.

Diver, C., A. E. Boycott, and S. Garstang. Inheritance of Inverse Symmetry. *J. Genetics, 15*, 113-200, 1925.

Eakin, R. M. Regional determination in the development of the trout. *Arch. f. Entw-mech., 139*, 274-281, 1939.

Fankhauser, G. Cytological studies on egg fragments of the salamander Triton. IV. The cleavage of egg fragments without the egg nucleus. *J. Exp. Zool., 67*, 349-394, 1934.

Gillette, R. Cell number and cell size in the ectoderm during neurulation (Amblystoma maculatum). *J. Exp. Zool., 96*, 201-222, 1944.

Hadorn, E. Übertragung von Artmerkmalen durch das entkernte Eiplasma beim merogonischen Triton-Bastard palmatus-Plasma X cristatus-Kern. *Verh. Dtsch. Zool. Ges., 1936*, 97-104, 1936.

Hall, E. K. Regional differences in the action of the organization center. *Arch. f. Entw-mech., 135*, 671-688, 1936.

Harrison, R. G. The effect of reversing the mediolateral or transverse axis of the forelimb bud in the Salamander embryo (Amblystoma punctatum Linn.) *Arch. f. Entw-mech., 106*, 469-502, 1925.

Harrison, R. G. Cellular differentiation and internal environment. *A.A.A.S., publ. no. 14*, 77-97, 1940.

Harrison, R. G. Relations of symmetry in the developing embryo. *Trans. Conn. Acad. Arts and Sciences, 36*, 277-330, 1945.

Harvey, E. B. Parthenogenetic merogony or cleavage without nuclei in Arbacia punctulata. *Biol. Bull., 71*, 101-121, 1936.

Harvey, E. B. Structure and development of the clear quarter of the Arbacia punctulata egg. *J. Exp. Zool., 102*, 253-275, 1946.

Heatley, N. G. The distribution of glycogen in the regions of the Amphibian gastrula; with a method for the microdetermination of glycogen. *Bioch. Jour., 29*, 2568-2572, 1935.

Heatley, N. G., and P. E. Lindahl. Studies on the nature of the amphibian organization centre. V. The distribution and nature of glycogen in the amphibian embryo. *Proc. Roy. Soc. London, (B) 122*, 395-402, 1937.

His, W. *Unsere Körperform und des physiologische Problem ihrer Enstehung.* Leipzig, F. C. W. Vogel, 1874.

Holtfreter, J. Gewebeaffinität, ein Mittel der embryonalen Formbildung. *Arch. exper. Zellforschg., 23,* 169-209, 1939.

Holtfreter, J. A study of the mechanics of gastrulation, part I. *J. Exp. Zool., 94,* 261-318, 1943a.

Holtfreter, J. Properties and functions of the surface coat in amphibian embryos. *J. Exp. Zool., 93,* 251-323, 1943b.

Holtfreter, J. Neural differentiation of ectoderm through exposure to saline solution. *J. Exp. Zool., 95,* 307-343, 1944a.

Holtfreter, J. A study of the mechanics of gastrulation, part II. *J. Exp. Zool., 95,* 171-212, 1944b.

Lewis, W. H., and E. C. Roosen-Runge. The formation of the blastodisc in the egg of the zebra fish, Brachydanio rerio, illustrated with motion pictures. *Anat. Rec., 84,* 13-14, 1942.

Lillie, F. R. Embryonic segregation and its role in the life history. *Arch. f. Entw-mech., 118,* 499-533, 1929.

Luther, W. Potenzprüfungen an isolierten Teilstücken der Forellenkeimscheibe. *Arch. f. Entw-mech., 135,* 359-383, 1936.

Luther, W. Transplantations und Defektversuche am Organisationszentrum der Forellenkeimscheibe. *Arch. f. Entw-mech., 137,* 404-424, 1937.

Morgan, T. H. Experimental studies on the teleost eggs. *Anat. Anz., 8,* 803-814, 1893.

Needham, J. *Biochemistry and Morphogenesis.* Cambridge University Press, 1942.

Nicholas, J. S. Blastulation, its role in pregastrular organization in Amblystoma punctatum. *J. Exp. Zool., 100,* 265-299, 1945.

Nieuwkoop, P. D. Experimental investigations on the origin and determination of the germ cells and on the development of the lateral plates and germ ridges in Urodeles. *Archives Néerlandaises de Zoologie., 8,* 1-205, 1946.

Oppenheimer, J. Experiments on early developing stages of Fundulus. *Proc. Natl. Acad. Sci. U.S., 20,* 536-538, 1934.

Oppenheimer, J. Processes of localization in developing Fundulus. *Proc. Natl. Acad. Sci. U.S., 21,* 551-553, 1935.

Oppenheimer, J. The development of isolated blastoderms of Fundulus heteroclitus. *J. Exp. Zool., 72,* 247-269, 1936a.

Oppenheimer, J. Processes of localization in developing Fundulus. *J. Exp. Zool., 73,* 405-444, 1936b.

Pasteels, J. La gastrulation et la répartition des territoires dans la moitié dorsale du blastodisque de Truite (Salmo iridaeus). *C. R. Soc. Biol., 113,* 425-428, 1933.

Pasteels, J. Répartition des territoires et mouvements morphogénétiques de la gastrulation de l'oeuf de truite (Salmo iridaeus). *C. R. Ass. Anat. Bruxelles, 1934,* 451-458, 1934.

Pasteels, J. Etudes sur la gastrulation des vertébrés méroblastiques. III. Oiseaux. *Arch. de Biol., 48,* 382-463, 1936.

Poulson, D. Chromosomal control of embryogenesis in Drosophila. *Am. Nat.*, *69*, 340-363, 1945.

Robertson, G. G. An analysis of the development of homozygous yellow mouse embryos. *J. Exp. Zool.*, *89*, 197-231, 1942.

Rose, S. M. Methods of initiating limb regeneration in adult Anura. *J. Exp. Zool.*, *95*, 149-170, 1944.

Rudnick, D. Early history and mechanics of the chick blastoderm. *Quart. Rev. Biol.*, *19*, 187-212, 1944.

Sawyer, C. H. Cholinesterase and the behavior problem in Amblystoma. I. The relationship between the development of the enzyme and early motility. *J. Exp. Zool.*, *92*, 1-11, 1943a.

Sawyer, C. H. Cholinesterase and the behavior problem in Amblystoma. II. The effects of inhibiting cholinesterase. *J. Exp. Zool.*, *92*, 11-29, 1943b.

Sawyer, C. H. Cholinesterase and the behavior problem in Amblystoma. III. The distribution of cholinesterase in nerve and muscle throughout development. *J. Exp. Zool.*, *94*, 1-12, 1943c.

Sawyer, C. H. Cholinesterase and the behavior problem in Amblystoma. IV. Cholinesterase in nerveless muscle. *J. Exp. Zool.*, *94*, 12-31, 1943d.

Schechtman, A. M. Unipolar ingression in Triturus torosus. *Univ. of Calif. Publ. Zool.*, *39*, 303-307, 1934.

Schotté, O. E. Influence de la section tardive des nerfs sur les pattes de Tritons en régénération. *C. R. Soc. Phys. et Hist. nat. Genève.*, *40*, 86-88, 1923.

Schotté, O. E. Système nerveux et régénération chez le Triton. *Rev. Suisse de Zool.*, *33*, 1-211, 1926.

Schotté, O. E., and E. G. Butler. Morphological effects of denervation and amputation of limbs in urodele larvae. *J. Exp. Zool.*, *87*, 279-322, 1941.

Schotté, O. E., and E. G. Butler. Phases in regeneration of the urodele limb and their dependence upon the nervous system. *J. Exp. Zool.*, *97*, 95-121, 1944.

Schrödinger, E. *What is Life? The physical aspect of the living cell.* Cambridge University Press, 1945.

Singer, M. The nervous system and regeneration of the forelimb of adult Triturus. I. The role of the sympathetics. *J. Exp. Zool.*, *90*, 377-399, 1942.

Singer, M. The nervous system and regeneration of the forelimb of adult Triturus. II. The role of the sensory supply. *J. Exp. Zool.*, *92*, 297-315, 1943.

Singer, M. The nervous system and regeneration of the forelimb of adult Triturus. III. The role of the motor supply, including a note on the anatomy of the brachial spinal nerve roots. *J. Exp. Zool.*, *98*, 1-21, 1945.

Singer, M. The nervous system and regeneration of the forelimb of adult Triturus. IV. The stimulating action of a regenerated motor supply. *J. Exp. Zool.*, *101*, 221-240, 1946a.

Singer, M. The nervous system and regeneration of the forelimb of adult

Triturus. V. The influence of number of nerve fibers, including a quantitative study of limb innervation. *J. Exp. Zool.*, *101*, 299-331, 1946b.

Singer, M. The nervous system and regeneration of the forelimb of adult Triturus. VI. A further study of the importance of nerve number, including quanitative measurements of limb innervation. *J. Exp. Zool.*, *104*, 223-250, 1947a.

Singer, M. The nervous system and regeneration of the forelimb of adult Triturus. VII. The relation between number of nerve fibers and surface area of amputation. *J. Exp. Zool.*, *104*, 251-265, 1947b.

Spemann, H., F. G. Fischer, and E. Wehmeier. Zur Kenntnis der Induktionsmittel in der Embryonalentwicklung. *Naturwiss.*, *21*, 518, 1933.

Spratt, N. T., Jr. Formation of the primitive streak in the explanted chick blastoderm marked with carbon particles. *J. Exp. Zool.*, *103*, 259-304, 1946.

Stableford, L. The potency of the isolated vegetal hemisphere (presumptive endoderm) of the blastula of Amblystoma punctatum. *Anat. Rec.*, *75*, abs. 35-36, 1939.

Tung, T. C., C. Y. Chang, and Y. F. Y. Tung. Experiments on the developmental potencies of blastoderms and fragments of teleostean eggs separated latitudinally. *Proc. Zool. Soc. Lond.*, *115*, 175-188, 1945.

Vogt, W. Gestaltungsanalyse am Amphibienkeim mit örtlicher Vitalfärbung. Vorwort über Wege und Ziele. I. Teil: Methodik und Wirkungsweise der örtlichen Vitalfärbung mit Agar als Farbträger. *Arch. f. Entw-mech.*, *106*, 542-610, 1925.

Vogt, W. Gestaltungsanalyse am Amphibienkeim mit örtlicher Vitalfarbung. II. Teil. Gastrulation und Mesodermbildung bei Urodelen und Anuren. *Arch. f. Entw-mech.*, *120*, 384-706, 1929.

Woerdeman, M. W. Über den Glykogenstoffwechsel des Organisationszentrums in der Amphibiengastrula. *Proc. Acad. Sci. Amsterdam*, *36*, 189-193, 1933.

Yamamoto, T. Studies on the rhythmical movements of the early embryos of Oryzias latipes. I. General description. *J. Fac. Sci., Imp. Univ. Tokyo*, *2*, 147-152, 1931a.

Yamamoto, T. Studies on the rhythmical movements of the early embryos of Oryzias latipes. II. Relation between temperature and the frequency of the rhythmical contractions. *J. Fac. Sci., Imp. Univ. Tokyo*, *2*, 153-162, 1931b.

Yamamoto, T. Studies on the rhythmical movements of the early embryos of Oryzias latipes. III. Temperature and the amplitude of the contraction waves. *J. Fac. Sci., Imp. Univ. Tokyo*, *3*, 105-110, 1933a.

Yamamoto, T. Studies on the rhythmical movements of the early embryos of Oryzias latipes. IV. Temperature constants for the velocity of the wave and for the pause. *J. Fac. Sci., Imp. Univ. Tokyo*, *3*, 111-117, 1933b.

Yamamoto, T. Studies on the rhythmical movements of the early embryos of Oryzias latipes. V. The action of electrolytes and osmotic pressure. *J. Fac. Sci., Imp. Univ. Tokyo, 3,* 287-299, 1934.

Yamamoto, T. Studies on the rhythmical movements of the early embryos of Oryzias latipes. VI. The action of hydrogen ion concentration. *J. Fac. Sci., Imp. Univ. Tokyo, 4,* 221-232, 1936a.

Yamamoto, T. Studies on the rhythmical movements of the early embryos of Oryzias latipes. VII. Anaerobic movements and oxidation-reduction potential of the egg limiting the rhythmical movements. *J. Fac. Sci., Imp. Univ. Tokyo, 4,* 233-247, 1936b.

Yamamoto, T. Studies on the rhythmical movements of the early embryos of Oryzias latipes. VIII. The effect of series of carbamates. *J. Fac. Sci., Imp. Univ. Tokyo, 5,* 37-49, 1938a.

Yamamoto, T. Contractile movement of the egg of a bony fish Salanx microdon. *Proc. Imp. Acad. Tokyo, 14,* 149-151, 1938b.

Yamamoto, T. On the distribution of temperature constants in Oryzias latipes. *Proc. Imp. Acad. Tokyo, 14,* 393-395, 1938c.

IX. NEOPLASTIC ABNORMAL GROWTH

BY C. P. RHOADS[1]

ABNORMAL growth is so broad a subject that the limits within which we are to use the term here must be precisely defined before it is discussed. The purpose of the consideration herewith presented is to explore the stockpile of possibly unrelated factual material of a fundamental biological nature for ideas and principles which may make more effective the investigative attack on neoplastic disease. This kind of abnormal growth is one which has no function in, or integration with, any orderly cellular structure and, more than this autonomy, it has a positive, malignant capacity to injure normal tissue. It is analogous to the predatory animal in the domesticated herd.

UNIQUE CHARACTERISTICS OF THE CANCER CELL

Extensive studies in progress are designed to provide new information about the nature of the cancer cell, a most important problem in the field of abnormal growth. Though repetitious, it is relevant to state once more that this cell is characterized by unique properties and is, supposedly, the repository of secrets which, if known, would make simple the solution of the whole great problem of neoplastic disease. Many investigators are dissatisfied with their limited knowledge of the mechanism which provides such a destructive and uncontrolled structure with its power, and some wish particularly to understand in greater detail how the constituent cells differ in their constitution and function from their normal precursors. Not only is this question of fundamental interest, but also, were adequate information concerning it at hand, a direct therapeutic attack on neoplastic disease conceivably could be developed.

A histological preparation of cancer tissue in contrast to its background of normal cellular structure provides ample evidence to establish the unusual capacities of the cancer cell. The very ability of the pathologist to diagnose a neoplasm from the forms and staining reactions of its components is proof enough (1) that they are in general strikingly abnormal. The cells are not only distinctive in form, but also show unique competitive powers. The morphology of a gastric wall infiltrated with cancer, or of a cervix uteri similarly involved, demonstrates adequately

[1] Director, The Memorial Hospital for the Treatment of Cancer and Allied Diseases, and the Sloan-Kettering Institute for Cancer Research. Professor of Pathology, Cornell University Medical College, New York.

that the neoplastic cell possesses special privilege and a profound biological advantage over its normal fellows. It apparently can carry out unusual reactions as shown by its abilities to invade, distort, and destroy normal tissue, to survive in the midst of violent inflammatory reactions, and to withstand invasion by microorganisms. It goes its own way—it has the property of autonomy and in many instances something more.

A simple quantitative increase in the rate of growth of cells, associated with a decrease in their ability to differentiate, sometimes has been advanced as a generalization sufficient to explain the unique properties of neoplastic tissue. It is desirable to establish at the outset how sound this view is, to decide whether it is sufficiently inclusive to comply with all of the facts, and to ascertain to what extent it can be used as a basis for planning new experiments, more conclusive than the ones now at hand.

The fact that some cancer cells multiply rapidly is proved by measurement of their metabolic processes. Rapid growth is not a distinctive quality, however, since certain normal cells, devoid of autonomy, multiply even more rapidly, as shown by Brues (2). The evidence for the limited differentiation of cancer cells is adequate, but also, like rapid growth, not uniformly present. In some instances, certainly, little evidence for the existence of any normal function can be detected, and in essentially all, the cellular activities seem to be less specialized than are those of normal cells of similar types. In the vast majority of neoplasms, however, some evidence of differentiated form and function is at hand and it may be very impressive. Hyperthyroidism due to functioning thyroid carcinoma in the absence of the thyroid gland (3) and the production of hormones by tumors of other endocrine organs provide satisfactory proof that highly specialized activities may continue even though the cells responsible for them clearly possess wholly abnormal malignant qualities of invasiveness and destructiveness. Clearly, something more than a quantitative change in the rates of growth and differentiation is required to convey the property of neoplastic growth.

The analogy is frequently drawn between cancer and embryonic cells on the grounds that both grow rapidly and are little differentiated. The embryonic cell certainly is not a neoplastic structure, however, nor is it likely to become one without very considerable alteration. It has no autonomy and lives dependent upon its associates. The cancer cell is far more competent, aggressive, and malignant, and less susceptible to normal control and to orderly differentiation than is its normal embryonic progenitor. These qualities must depend upon some profound deviation from the normal structure or function, or both.

The distinctive characteristics of neoplastic growth are as striking as those of the cells of the mutant bacterial colony described by Haddow (4): "The generation of a vigorously growing strain of bacterial cells in a senescent colony presents a striking resemblance to the emergence of a malignant variant in the cells of aging animal tissues." The comparison is supported by many examples of adaptive variations in bacteria. Here the cells of secondary colonies have been shown functionally and structurally to be unlike the parents from which they arose, due to the fact that they can metabolize constituents of the medium which the progenitors are unable to attack at all. For lack of this competence the parent cells, even though young and actively reproducing, are handicapped in their ability to carry on, with the substrate available, the metabolic activities required for active proliferation. The variant, analogous to the malignant cell, more versatile and perhaps better equipped biochemically, can thrive under circumstances which are intolerable to its less adaptable parent.

The Control of Cellular Growth and Differentiation

The role in cancer of a loss of the normal control of growth and differentiation should be considered, whether or not we accept as correct the conclusion that the constituent cell must involve something more than a simple quantitative deviation of these factors from normal. There are two possibilities: a wild growth, like that of cancer, might be due to a local lack or a modification of some general control of development and function; or, more possibly, the general force could remain unaltered but be powerless to hold in restraint cells which have assumed properties different from those of normal cells' properties through which control is exerted under ordinary circumstances. Clearly, if a generally applicable force has been modified to cause or to allow cancer to develop, the modification must be one which affects particularly a single specific group of cells. It seems more probable that the cancer tissue has some marked structural or functional difference from the normal, one which should be detectable if searched for adequately. It is of the utmost importance that any such difference be sought out and defined with the greatest precision possible.

Certain controls of differentiation exist which, if disordered, may concern the variation which is characteristic of cancer. One example is found in the fact that when tissues composed of undifferentiated cells come into contact, they unite. This is seen in wound healing, during embryological development, and with neoplastic tissue. The temporary union of undifferentiated cells and their subsequent permanent junction is an important architectural process in embryology (5). The terminal part of the floor of the penile urethra, for example, is formed by the union of the lateral walls of the urethral groove, and in the normal development of the embryo is replaced by ordinary firm connective tissue. Gonadal hormones can cause the two uniting epithelial edges to become keratinized so that union is impossible and hypospadias results. The squamous cancer of the uterus induced in mice by Allen and Gardner (6) provides a notable example of wholly abnormal differentiation caused by an extracellular agent. The whole vast field of cancer production by hormones is relevant to this question and makes available ample evidence of the susceptibility of cells to general controls. It would be desirable to have additional data concerning the effects of the hormones and the carcinogenic chemicals on embryological development. We lack precise comparative observations of their effects in causing differentiation and possibly dedifferentiation of similar material susceptible to precise study. Reference to Needham (7) is desirable on this point.

The future may well see experiments designed to test more exactly the flexibility of function of cells at various stages during differentiation. It is not impossible that the assumption of adult differentiated form involves a constant succession of losses of genes, or controlling elements, with consequent loss of ability to carry out a variety of functions coincident with concentration on a comparatively small number. This emphasis on certain limited functions could well give the distinctive characteristics and abilities possessed by highly specialized tissue. The constant tendency toward decrease in the number of functions capable of being executed would, of course, lead to a progressive handicap to the cell in surviving in a standard environment and in time to the very impossibility of doing so. This progressive change would eventually mean the death of the cell, and indeed, this is precisely the fate of differentiated tissue. Is it too fanciful to believe that the endocrine glands are accessory organs created to maintain, in their terminal, highly differentiated phases, the metabolic activity of cells which are necessary for only a part of life's span? Certainly when these glands are extirpated, specific highly differentiated cellular systems are put at a great disadvantage and die, or at least become inactive.

The fact is remarkable that, to date, in spite of the enormously extensive studies which have been made of neoplastic tissue and the obvious morphological contrasts between normal and cancer cells, no qualitative chemical differences have been established conclusively. A considerable series of studies culminating in the recent work of Friedewald (8), of Kidd (9, 10, 11), and of Green (12, 13) provide certain immunologic evidence for the presence of abnormal constituent proteins in the cancer tissue of animals. Very great extension of the work is required, since in few instances described did the tumor studied arise in animals homozygous with the ones in which antibodies were induced. The measurements of the enzymatic activities of neoplastic tissue instituted by the Warburg school (14) and so effectively continued by the investigators of the National Cancer Institute (15), the McArdle laboratory (16), by Salter (17) and others have not revealed completely uniform qualitative differences between normal and neoplastic enzyme function. Can it be that new enzymes have not been found because of the tendency of investigators to examine cancer tissue for the activity of normal systems against normal substrates, rather than to search empirically for new metabolic pathways? Differences of structure and function must exist, however, between normal and cancer cells, even though they have not been satisfactorily defined. Perhaps a new approach can be developed from a consideration of the possible origins of differences between cells.

Cell Characteristics as a Function of Genes

The induction of changes in cells is frequently susceptible to analysis in accord with the recognized principles of morphogenesis, under which differentiation and growth are governed by genes. Whereas this control of the patterns of living organisms is well established, it is not entirely clear how tissues and organs become differentiated into forms which are completely distinct one from the other, when supposedly all the component cells, being derived by mitotic divisions from a single fertilized egg, have identical sets of genes. If genes are the sole regulating agents, when a cell changes as it does in differentiating, the alteration should follow and be the result of a mutation of the genes. It is likely, however, that there exists in the cytoplasm of some cells a second group of controlling factors which are distinct from the genes of the chromosome thread, but which may depend on the genes for their origin and to some extent for their continued growth. Conceivably both genic and extragenic regulatory mechanisms operate coincidentally.

The concept of governing factors in the cytoplasm, though almost as old as cytology, has recently been supported by new and limited but convincing evidence reviewed by Haddow (18). It is a subject of great importance to the cancer problem because it provides a principle of morphogenesis which has been involved to explain the transmission of some unusual but actively studied examples of neoplastic disease by cell-free agents (viruses) in part derived from avian tumors. Some have assumed that these agents must be cytoplasmic entirely, and require little consideration of a genic origin, though adequate justification or even need for this assumption is hard to establish.

Satisfactory data attest to the view, however, that a very considerable and important part of the form and function of sexual cells, as well as of vegetative forms, is controlled by genes without the participation of any non-genic or cytoplasmic factor. Hence, since a normal adult cell differs profoundly from its primitive undifferentiated precursor, the genes of the two forms should differ also. In support of this possibility is the fact that differentiated cells contrast sharply with their primitive relatives in their susceptibility to injury by one agent, x-rays, known to have a profound effect on genes (19). Similarly, since the form and function of the cancer cell differ distinctly from those of the normal cell, the genes of these two cell types also should be different. Morphological grounds to support this view are at hand in a long series of studies, of which the most recent are those of Biesele (20). Although obvious abnormalities of chromosomes are not present in every cell believed to be neoplastic, they frequently can be demonstrated. Furthermore, it is clear from the results of experiments to be discussed later (21) that genic changes adequate to cause profound cellular alterations may exist without any morphological evidence of their presence.

As a basis for the concept that the progressive steps in cellular differentiation, normal or abnormal, can be controlled by genes, proof must be advanced that genes can mutate spontaneously in a regular fashion, since only in this way could they control the changes in structure and activity associated with orderly development. Demerec (22) has presented adequate evidence that genes vary greatly in their stability. Some genes of Drosophila, for example, mutate rarely or not at all, whereas others, and these are numerous, particularly in plants, mutate hundreds of times in the development of a single individual. Plough (23) has shown that the spontaneous lethal mutation rate in two of the three large chromosomes of Drosophila is in direct proportion to temperature and that the mutations so induced may subsequently disappear. Furthermore, the existence of genes which affect in turn the spontaneous mutability of other genes has been proven by Rhoades (24).

Once the fundamental instability of genes has been established, if we are to invoke them as regulators of differentiation, it is necessary to specify what changes of cells are involved in that process and to establish that they are alterations of constituents which are dependent upon specific genes for their existence. If this can be done, and if we are satisfied that those particular genes exist in mammalian cells, then, pending the submission of strong evidence concerning the regulatory function of cytoplasmic factors, it is not unreasonable to consider the possibility that orderly, genic controlled variation is adequate to provide a basis for development, either orderly or disorderly. Beadle states: "Cells of identical genotypes travel many divergent paths in the process of differentiation of the individual organism" (25). The different paths involve the acquisition of structural and functional characteristics which concern, particularly, the protein components of the cell. Satisfactory proof is available that organs of special function are constructed of different and characteristic proteins; the composition of the skin, for example, is demonstrably quite different from that of other tissues. Functional differences between cells depend upon kinetically active protein enzymes, and examples of their specificities in differentiated tissues are almost numberless. With the special proteins and enzymes established as characteristic of cellular differentiation, proof is required that the production of these constituents is controlled by genes.

Ample demonstration that genes control the constitution of tissue proteins is found in the antigenic specificities of tissues as compliant with the principles of heredity. From Landsteiner's (26) contributions concerning the blood groups of man, through the more recent studies of Wiener (27) on the Rh factor, a mass of evidence is available. Particularly striking is the fact that a 1:1 gene-to-antigen relationship has been shown to exist in experimental material by Tatum (28), a fact which suggests that the antigens are direct products of the genes, or even further, that the antigen is "the copy of the gene." The thesis of the controlled production of antigens by genes is further supported by immunological studies referred to later, in which genes apparently have been altered by use of antisera specifically directed against the antigen, or gene product.

Striking evidence that enzyme production is also under the control of genes is to be found in the experiments of Beadle and his associates, well summarized in a recent publication (25). Elaborate studies of Neurospora have proven beyond any question that the enzymatic activities of those cells and, by inference, cells of any type, are formed by genes. Beadle states as follows: "Genes evidently determine the final potentialities of differentiation in a given organism. Whereas the initiation of the orderly process may be through the environment, whether or not a given reaction . . . can take place is gene controlled." An example is at hand in the MacDowell (29) dwarf mice, in which the normal gene concerned is responsible for the development of pituitary eosinophils. Without this gene the pituitaries of the dwarf animals lack the eosinophilic component which, if supplied artificially, restores growth. Because of this lack the growth extent and rate of all organs is enormously reduced.

If the control of normal cellular development and differentiation by genes is indicated by the evidence for genic regulation of protein and enzyme production as well as for the existence of spontaneous variations in genes, and if a similar mechanism is to be invoked in the control of *abnormal* differentiation, as of the cancer cell, it becomes necessary to establish the possibility of *abnormal* mutations. Mammalian cells are vegetative and hence not susceptible to genetic study by standard methods. For this reason, the last step possible in a correlation of abnormal variation with cancer is to show that the agents which produce cancer cells also cause mutations in sexual forms. Mutations can be produced artificially by certain physical and chemical agents, and apparently by proteins with the properties of antisera. The physical agents include x-rays, gamma rays, neutrons, and ultraviolet light, and a mass of evidence proves that all of these are carcinogenic.

The Artificial Production of Gene Changes and Cancer by Physical Agents

The studies of mutations as caused by x-rays are classic. Chromosome aberrations are said to result from x-rays more frequently than they occur spontaneously (30), but true mutations of the spontaneous types are also caused. Many have been proven to be due to the removal of small segments of chromosomes followed by junction of the broken ends. Some cannot be due to such a chromosome deficiency, however, and rather must represent an actual change in the composition of the gene, since back mutations occur. The mutations appear to start as changes in individual atoms, in consequence of which, by a chain of reactions, the mutation results. The question of whether gene mutation can be regarded as minute rearrangements of a size too small to be seen cytologically has been reviewed by Muller (19).

Glucksmann reports that under certain conditions the process of differentiation of mammalian embryonic cells is promoted by irradiation (31), and that their sensitivity to injury decreases as differentiation proceeds. Can it be that the differentiated cell is no longer sensitive to certain dosage of x-rays because many of the points on the chromosome which are particularly vulnerable to this agent have previously altered spontaneously? Adequate data are at present not at hand to prove that x-ray effects are not additive and that they cause variations which are resistant to the agent. Work of this type is in progress and the results are said to indicate that resistance does develop.

The experiments of Beadle (25) prove completely that defects in certain biosyntheses can be produced in Neurospora by irradiation with x-rays with resultant mutations of specific genes. The growth factor requirements of the mutated organism result from the induced inability to manufacture essential substances from simpler constituents. These requirements are characteristic of a given strain and are transmitted to each descendant cell, unless back mutation occurs. Gray and Tatum (32), furthermore, proved that x-ray treatment of two species of bacteria also produced strains characterized by their inability to carry out specific biochemical reactions and suggested that biosyntheses in bacteria, like those of Neurospora, are controlled by specific genes.

The evidence indicates that a single molecular change affects a single gene of sexual cells. A similar effect in vegetative cells is suggested by the fact that in the case of both viruses and bacteria a single atom ionized can kill. Lea (33) has shown that this atom cannot be one located just anywhere; it must be in a sensitive component, just as in cells which have demonstrable genes. Gowen (34) compared the mutation rates induced by x-rays in Drosophila, tobacco mosaic virus and the bacteria causing corn wilt disease. The frequency of variation was increased at the same rate in all three, a fact which again suggests a similar structure controlling mutation in both sexual and vegetative cells. These observations provide examples of the similarity between the variations induced by the same means in sexual (mold) and in vegetative (bacterial) cells. Since genes are involved in the former, it is reasonable to assume the participation of genes or gene-like substances in the latter. This conclusion is, of course, not susceptible to absolute proof, a point conceded by the authors. Furthermore, it is not possible to prove whether the mutation has occurred in hypothetical nuclear genes or in hypothetical cytoplasmic determiners or plasmagenes.

It is entertaining to speculate on cellular differentiation as a progressive series of losses of function, like those undergone by mutated Neurospora, by which the cell retains the ability to exercise a certain number of special functions while, through mutation, the abilities are lost to carry out certain other activities to completion. The formation of pigment, for example, or the accumulation of any metabolite, could be simply the result of loss of ability to carry out completely a series of metabolic steps.

The simple gene-hit theory of x-ray effects has been brought into question by the experiments reported by Dale (35) on the changes induced by x-rays of two enzymes in solution. These suggest that the molecules of the solvent are activated by the absorption of radiation and transfer their energy to the dissolved protein. The fact that enzymes can be so inactivated suggests the existence of an intracellular effect of radiation on enzymes somewhat more complex than is explainable by the simple gene-hit theory. The work on Neurospora proves that a direct relationship exists between genes and enzymes. Hence the Dale mechanism may simply prove the existence of an indirect route, for the x-ray effect, through enzymes to genes. Lea, Smith, Holmes, and Markham (36) have provided further evidence on this point by showing that virus inactivation by radiation may be either direct, independent of concentration of the solution, or indirect in dilute solution.

The ability of x-rays to cause cancer has been amply established by tragic experience among radiologists and requires little documentation. Particularly detailed confirmation is found in the recent experimental studies of Henshaw (37) on the production of leukemia in mice by irradiation. Furth (38) has reported that granulosa cell tumors of the ovary can be similarly caused in mice. A most interesting observation on the neoplastic effects of x-rays on man is that of March (39), who finds that the incidence of leukemia in radiologists is more than ten times that among physicians who are not radiologists.

Ultraviolet light, also a cause of cancer, seems to be somewhat less effective than are x-rays in the induction of mutation. The effects of the two agents on maize were carefully compared by Stadler (40), who found certain differences. An explanation may be at hand in the length of the time interval between the break and the separation of the chromosome fragments when the mutation is due to ultraviolet. The striking effect of certain wave lengths of the ultraviolet is in sharp contrast to the apparent absence of wave length effect in the x-ray range. Beadle (25) states: "The curve of effectiveness of ultraviolet in producing mutations against the wave length effect is sufficiently similar to the absorption spectrum of nucleic acid over the same wave length as to support strongly the view that this substance is closely associated with genes if not a component of them." The communications of Hollaender (41) should be consulted on this point. This investigator points out that on living cells the 2600 A region of ultraviolet is the most productive of genetic effects. With Greenstein (42), he irradiated sodium thymonucleate in water and observed that a decrease in streaming double birefringence and viscosity occurred. A depolymerization was postulated. Hollaender concluded that there was a primary absorption of the ultraviolet by nucleic acid followed by the breakdown of the protein and the disruption of the relation between the nucleic acid and the protein.

It is important in this regard to consider the observations of Stanley on variants of tobacco mosaic virus (43). Differences in the amount of one or more amino acids, the presence of a new amino acid, and the complete elimination of one amino acid were found. This indicates again that, in variation, changes in protein composition may be quite as important as effects confined solely to nucleic acid. Work of this type on mammalian cells is urgently needed, and certain studies are in progress under Brown in the Memorial Hospital laboratories.

A particularly interesting observation, and one in keeping with Stanley's results, is that of MacKenzie and Muller (44), who induced mutations of Drosophila by wave lengths from 2800 to 3650 A. The effect was especially marked at 3100 A, with a second maximum at 2280 A, presumably a result of the absorption of energy by the protein component. It would be interesting to explore this point further with other types of organisms.

Gray and Tatum (32) report failure in their efforts to cause variations analogous to mutations in sexual forms (Neurospora) in bacteria by ultraviolet, but Beadle and Tatum (90) were successful in the case of Neurospora, and rather more mutant strains were produced by ultraviolet than by x-rays. Certain differences are reported between the effects of the two agents. All but one of the mutations resulting in an inability to synthesize thiamine were caused by x-rays and, in contrast, those interfering with inositol and choline formation followed ultraviolet exposure. This point is a particularly intriguing one since it suggests the possibility that certain genes are more sensitive than others to one specific physical agent.

Blum (46) has reviewed the effects of ultraviolet light in great detail. It is clear from his work that the wave length range responsible for mutations and the absorption spectrum of nucleic acid in the ultraviolet correlate closely with the effects of the light in causing cancer (as well as erythema) of the skin in experimental animals. Here again is evidence that a very specific agent, a narrow wave length range of ultraviolet, clearly causative of mutations (or back mutations) in genetically controlled material, is also active in inducing the change from the normal to the neoplastic cell.

The decreasing sensitivity to radiation shown by more differentiated cells is striking and lends interest to any means by which the sensitivity to mutation can be reduced. Swanson (47) has shown that if ultraviolet is used followed by x-rays, the effects of the latter on chromosome aberrations are partly suppressed. These findings suggest that the structures in the chromosome are not equally sensitive and that certain ones are particularly susceptible to change induced by a specific physical agent. This cannot be a rule, however, since a strain of Neurospora has been reported with two different mutated genes, one the effect of x-rays and the other of ultraviolet (48).

Mutations Caused by Chemical Agents

The induction of inherited changes, but not of cancer, by antisera has been reported. Guyer and Smith (49) injected an antiserum to rabbit lens protein into pregnant rabbits and reported that transmissible ocular defects occurred in the offspring. Hyde (50) is stated to have confirmed these observations. Lens protein is the product of a gene and not a gene itself, but it may be that the modification of the product, an antigen, modifies the gene in turn. It is also possible, as specified by Sturtevant (50), that genes have chemical specificities corresponding to the antigens which they produce, and so are affected directly by specific antibodies. The work of Emerson (51) tends to confirm this assumption. He obtained from culture filtrates enzymes, described as extracellular and adaptive in character, against which immune sera were prepared and used to treat Neurospora. Resultant inherited changes in the ability of the mold to produce the enzymes are reported. It is possible that this principle provides an explanation of the results of Kidd (52) and of Green (13) in injuring cancer cells by antisera against them.

For a very considerable period chemical agents other than antisera which would cause gene mutation were actively sought, but clear proof of their existence was lacking. More recently, however, a group of compounds known as the halogenated tertiary alkyl amines (nitrogen mustards) has been studied. From the effects of these substances on Drosophila (53), on Neurospora (54), and on tobacco mosaic virus (55), it is clear that they are effective in causing mutations of classic types. It is interesting, furthermore, that they affect tissues *in vivo* very much as x-rays do. Inhibition of mitosis, injury to intestinal epithelium, damage to hematopoietic and lymphoid tissue, and necrosis of skin are all effects caused by both agents (56). Furthermore, the studies of Bodenstein (57) reveal that the nitrogen mustards, applied to growing tissue of the proper type and stage of development, stop cellular proliferation and cause extensive swelling, possibly an effort toward differentiation. This has also been shown by Karnofsky (58) in the Memorial Hospital laboratories to be true for neoplastic cells of one type, mouse sarcoma 180, grown on the chick embryo.

Another set of genetically active chemical agents are the carcinogenic substances of classic type containing a phenanthrene nucleus. When applied to tissue they cause profound alterations of cells which lead to self-reproducing alterations coincident with the assumption of neoplastic characteristics. Stowell (59) has shown that these changes are associated with alterations of nucleic acid content and Biesele (20) has demonstrated irregular alterations of chromosome structure. The metabolic and chemical abnormalities of the cancer cells induced by carcinogens have been so completely reviewed by Burk (60) and by the students of the Cowdry school (61) as to leave little doubt of their importance.

Unfortunately, few data are at hand on the use of carcinogenic chemical agents on material susceptible to genetic study. Beadle (25) states that tests with these substances have not resulted in mutations. Tatum (62) is said, however, to have made certain new observations which provide acceptable evidence that mutations can be induced in Neurospora by methyl-cholanthrene. Mottram (63) and Spencer and his associates (126) have reported striking changes in protozoa, and Strong (64) has recently published impressive studies on genetic changes in mice which have resulted from the administration of the same compound to young animals. Further work appears to be indicated.

A possible explanation of the difficulty in causing mutations in Neurospora by the chemical carcinogens is the fact that they are altered by contact with body cells and that the active compounds are quite as likely to be the intermediate metabolites as they are to be the original materials. Dobriner and Rhoads (65), Boyland (66), Jones (67), and Weigert (68) have all reported on studies of the metabolism of these substances. The very great insolubility of the original materials and the data which attest to the formation of highly labile intermediate products in the course of conversion support the view that the attempts to cause mutations may not have subjected the cells studied to exactly the compound which is active in the mammalian tissue.

The studies of Earle (125) are particularly important as they concern the alterations undergone by cells in the course of becoming malignant. He cultured mouse fibroblasts in the usual heterologous serum-chick embryo extract mixture. Subcultures were exposed for different periods to varying concentrations of methyl-cholanthrene, and the cells were inoculated into mice of the strain from which they had been derived. The cultures assumed characteristic growth characteristics and certain of them, after a rather narrow range of brief exposure to the carcinogen, were found to have neoplastic properties. This was also true of the mouse fibroblasts cultured for longer periods without methyl-cholanthrene.

The conclusion from this work is unavoidably that mouse fibroblasts grown for long periods under very abnormal conditions, particularly with a carcinogen, become sarcoma cells. Further details of the environmental circumstances to which the cells are exposed and the composition of the abnormal cultures will be anticipated.

As far as is known, no studies acceptable to the geneticist have been made on cells of the mutation-producing effect of the azo-carcinogens of the type of dimethylamino-azobenzene. In the case of these compounds much information is at hand from the studies of Kensler (69) and his associates, as well as from the Wisconsin school (70), concerning the pathways followed in the breakdown of the parent substance and the biological activities of the various derivatives. For example, dimethylamino-azobenzene is wholly inactive in inhibiting the metabolism of the tissue cells which have been studied manometrically in acute experiments. Certain derivatives, however, are exceedingly toxic. Suggestive evidence is at hand, furthermore, that other derivatives, not hitherto isolated, may also be biologically active. It is not sufficient to conclude that because the juxtaposition of a carcinogen, such as methyl-cholanthrene, and a cell susceptible to mutation, such as Neurospora, fails to give mutation, the chemical is incapable of doing so. It must be shown, before this conclusion is reached, that the cell has been exposed to all the derivatives of the carcinogen to which mammalian cells are exposed in the course of carcinogenesis.

Pure Speculation Concerning Normal and Abnormal Mutations

From the data it is clear that certain correlations exist between the production of mutations and the production of cancer. In general these correlations concern the ability of some agents to do both things, and the fact that the agents affect, in doing both, those constituents of cells which are composed of a material, nucleic acid, which is known to be involved in genes and in self-perpetuating alterations of vegetative bacterial material. Opposed to this concept of cancer as due to a discontinuous variation (mutation) are two considerations: first, the so-called "virus theory" by which the neoplasm is held to be due to an infection of the cell by a foreign parasite; and second, the fact that the cancer cell clearly is at a competitive advantage as compared to the normal cell, apparently able to carry out a vast number of processes not available to its more differentiated normal associate. Were the cancer cell a mutation in the accepted sense in which the term is used in describing the experiments on Neurospora, it should be less rather than more competent, since most mutations studied precisely have been shown to involve a *loss* rather than a *gain* of function. This fact requires a further consideration of development and of the possibility of back mutation.

Perhaps normal differentiation of cells is, after all, simply the result of mutations occurring normally at a predictable rate in an organized fashion with the continued growth only of those forms which have a superior adaptation to their environment. The word "mutation" as applied commonly to cancer should by this concept be "abnormal mutation," defined more precisely as an abnormal change occurring at an unpredictable rate and in a specific fashion. This is productive in large part of lethal effects or of pathological conditions including cells which are out of environmental control and so can compete to great advantage with their associates. Beadle (71) states, ". . . in present-day cellular organisms there exists a possible mechanism for acquiring totally new reactions. Occasionally through accident, one or more genes become duplicated, i.e., a small segment of a chromosome occurs twice in every set. But such a duplicate gene may undergo mutation in such a way that it directs the formation of an entirely new enzyme. If this new enzyme should happen to catalyze a reaction that improved the organism in competition with its relatives the new reaction would be retained."

The most dramatic experiments in support of this statement are those of Lindegren (72). He studied the enzymes which ferment melibiose and lactose as they are produced by yeasts of three strains and hybrids between them. The enzymes are adaptive in character and appear, in a genically qualified yeast strain, upon exposure to their respective substrates. All segregants from heterozygous diploids carrying a single pair of genes were able to adapt. Furthermore, two of the four adapted melibiose-fermenting cultures lost the function when dissimilated, adequate proof that the enzyme was transferred from the cytoplasm of the heterozygous hybrid maintained on melibiose to the cytoplasm of segregants which did not contain the gene. The enzyme was maintained in these segregants as long as the sugar was present. The evidence is good that the enzymes are self-perpetuating cytoplasmic entities, initiated by genes, with levels independent of the gene but dependent on interaction between the enzyme and the sugar.

Unfortunately, all the genetic considerations of cell differentiation as the result of mutations, either spontaneous or induced, are rendered unsatisfactory by the fact that our knowledge of gene action is all derived from organisms which reproduce sexually. Cancer cells are vegetative. Hence, if a consideration of variability and adaptation in other forms has some pertinence to the cancer problem, we must examine the facts regarding variations in other vegetative cells. Bacteria provide such examples, and reveal again the fact that those agents which cause mutations in sexual organisms susceptible to genetic study also cause inherited alterations, usually toward losses in function, in vegetative forms in which we have no way of demonstrating true genes, and are active in the production of cancer.

Bacterial Variation as Applicable to Abnormal Growth

Pure cultures of microorganisms, like the tissues of the body, even though they are derived from isolated single forms, are known not to consist entirely of identical individuals. Under standard conditions bacteria vary at a constant rate. If the variation provides the cell with an advantage in its environment, the variant continues to multiply and overgrows the cells of the type from which it arose. If it has derived no advantage from the variation, it is overgrown and dies. Modifications may be transient, and if so are usually a direct effect of environment and return to normal if the environmental alteration is removed; or they may be permanent, of a type for which no environmental cause can be found. It may be assumed that these permanent changes are analogous to the spontaneous mutations which occur in sexual forms. A third type of variation occurs during the growth cycle, and may be analogous to the changes leading to differentiation in more complex cells.

The changes which are transmitted by the altered cell to its progeny are termed discontinuous variations and include certain of the constitutive enzymes. The classical variation of this type is the unexplainable occurrence of lactose-fermenting forms of E. coli mutabile, normally a non-lactose-fermenting organism. It has been shown by Lewis (74) that all colonies of this organism generate about the same number of variant cells, but so few as to escape detection by conventional methods. The variation may occur in the absence of lactose.

Dubos (73) states that in bacterial cultures non-beneficial variations such as inadequate utilization of nutrients can be eliminated by overgrowth of the more vigorous original cells. This statement is an important one, since it should also apply to the more complex cells of tissues. However, normally it would not do so in the case of handicapping variations of differentiated tissue cells because, in contrast to movable bacteria, tissue components remain fixed in their relative positions. The differentiated cell, which perhaps has normally mutated, becomes specialized in its function, loses its adaptability, and has to compete only with similarly specialized cells adjacent to it. The most adult do not come in direct contact with the young and particularly vigorous cells, and so do not compete under the handicap imposed by the greater adaptability of primitive forms. Thus, in tissue, cells limited to new and specialized functions, such as seem to appear during differentiation, are protected in their existence unless there develop in their midst more competent and adaptable forms, such as neoplastic cells may be.

The development of abnormal cells in tissues can be caused experimentally by the application of certain agents as specified previously. All of these cause mutations, and all have a pronounced toxic or inhibitory effect on the normal cells with which they come in contact, with the result that the pathological forms, characteristic of neoplasm, appear. Perhaps analogous to this development of cancer following the prolonged intoxication of normal cells are the discontinuous variations of bacteria which yield abnormal cells characterized by their resistance to injurious agents. Antibacterial substances such as gentian violet and penicillin have been shown to cause the production of small resistant colony variants of staphylococci (75). Many species of bacteria have been exposed during their growth to immune sera with resultant changes in their antigenic (chemical) structure (76). These experiments must indicate, as discussed previously, that some factor which controls the self-reproducing character of bacteria, analogous to genes in the case of Neurospora, is changed as an effect of antisera, prepared not against the genes themselves, but against the material which they produce. It would be fascinating to know in what way, if any, the further sensitivity of bacteria to agents such as ultraviolet light at 2650 A, a wave length known to injure specifically the genes of sexual forms, is altered when a discontinuous variation due to the same agent has occurred. Evidence is becoming available that resistance to ultraviolet does follow exposure to it.

Whereas the evidence indicates that each bacterial organism has an inherent potentiality for producing a variant form once in a given number of divisions, it is difficult, as with tissue cells, to analyze these phenomena in the terms of classical genetics. To do so would of course require the ability to see and to describe alterations of structures in the vegetative cell which could be interpreted as having a function like the chromosomes and genes of sexual forms.

The studies of Lea (33) and of Gowen (34), mentioned previously, bear on this point. From this work it is clear that the use of a physical agent, x-rays, causes inherited alterations in viruses and bacteria similar to the ones caused in the more complicated types of higher organisms. In sexual forms the changes are known to be due to alterations of the genes. In both sexual and vegetative cells the changes are believed to follow hits on certain sensitive areas of the affected organism, areas which appear to be functionally quite analogous to the genes.

The transmutation of pneumococcal types, although it has been discussed repeatedly, deserves mention here. It can be depicted by a simple formula: Avirulent, unencapsulated Type II plus killed Type III equals virulent, capsulated Type III.

The reaction requires for transformation a competent culture. This culture may vary and result in non-encapsulated variants incapable of reacting to the transforming substance. It is clear that there exist a number of forms of unencapsulated variant pneumococci endowed with varying potentialities and physiological abilities. The study is best summarized by a direct quotation from Dubos (73a):

"The data obtained by chemical, enzymatic, and serological analyses together with the results of preliminary studies by electrophoresis, ultracentrifugation, and ultraviolet spectroscopy indicate that, within the limits of the methods, the active fraction contains no demonstrable protein, unbound lipid, or serologically reactive polysaccharide, and consists principally, if not solely, of a highly polymerized, viscous form of desoxyribonucleic acid. On the other hand, the Type III capsular substance, the synthesis of which is evoked by this transforming agent, consists chiefly of a non-nitrogenous polysaccharide constituted of glucose-glucuronic acid units linked in glycosidic union. The presence of the newly formed capsule containing this type-specific polysaccharide confers on the transformed cells all the distinguishing characteristics of Pneumococcus Type III. Thus, it is evident that the inducing substance and the substance produced in turn are chemically distinct and biologically specific in action and that both are requisite in determining the type specificity of the cell of which they form a part."

Information which is perhaps pertinent to a better understanding of the unique qualities of the cancer cell may be derived also from studies of the factors which make a bacterium pathogenic. The loss of the O type specific polysaccharide characteristic of the S→R variation of gram negative bacilli, like the loss of the specific polysaccharide of the pneumococcus, is associated with loss of virulence. The M substance of group A streptococci gives type-specificity and seems to be characteristic of all virulent strains. It is usually supposed that the disappearance of characteristic substances in bacterial variants is due to the fact that these substances are not produced. It may be, on the other hand, that in the variant cells the special substances are metabolically broken down by the acquisition of new functions. Although it is usually stated that the non-specific or undifferentiated bacteria have *lost* the ability to produce the O antigens or the capsular polysaccharide or the M protein, the possibility has not been ruled out that in the transformation from differentiated to undifferentiated cells they have become able to metabolize those substances further and hence do not accumulate them in recognizable form. Perhaps by this concept an analogy can be drawn between bacterial variation and the differentiation of mammalian cells. The assumption by tissue of special characters and powers coincident with maturation may actually represent discontinuous variations which involve the inability to carry certain reactions to completion, with the consequent accumulation of the products of these incomplete reactions.

Recent studies by Tatum (77) bear on this point. A mutant strain of Neurospora unable to synthesize nicotinic acid accumulates a special pigment, never formed by its progenitor, the wild type organism which can synthesize from nitrogen and sugar all of its components except biotin. The pigment is formed from an unused precursor of nicotinic acid. It is not the precursor itself, but rather a product of conversion of the precursor to pigment by an enzyme not demonstrable in the wild type mold. By a mutant blocking of one system, that for synthesizing nicotinic acid from the precursor, a new enzyme previously unused because unneeded, comes into play. The apparent appearance of a new function is largely the disappearance of an old one.

Concerning the abnormal functions of cells, particularly those of neoplastic tissue, a generally similar but abnormal process conceivably could have taken place with the production of variants possessing wholly unique abilities. For such a cell to survive requires that it not only be successful in competition with the normal cells in the company of which it finds itself, but also not susceptible to injury by whatever general mechanisms are possessed by the body for the control of cell growth.

Genetics and the Cancer Cell

The material presented provides a basis for a consideration of cellular differentiation, normal and abnormal, in terms of the physiology of simple forms. If differences between cells are genetic, they must be due to somatic mutation, the alteration of genes. No final proof of gene effects in the production of cancer is possible, since tissue cells are vegetative and not sexual, and since chromosome changes are not regularly demonstrable. Mutation is believed to be a possible cause because cancer can and does arise in localized areas, perhaps from single cells, and it results regularly from agents known to cause mutation. Since genes control specific steps in metabolism, they must have altered in the production of neoplastic cells of abnormal composition. Beyond this point of speculation we can do little more than argue from analogy. It may be useful, however, to set down what is known about how cancer cells differ from their normal prototypes.

The earliest experimental information concerning the cancer cell came from the demonstration that under certain circumstances it could be transplanted into otherwise normal hosts. This naturally required reference to the rules which apply to the transplantation of normal tissue from one animal to another. In general, an animal will not accept a graft containing a dominant allele which it itself does not have. The reverse, however, is not true, and dominant alleles in the host are without effect upon the acceptability of the graft. It is probable that these dominant genes are concerned with antigen formation. Lacking a gene competent to form an antigen, the foreign cell has no power to evoke an antibody response and so to subject itself to destruction. If the foreign cell or graft contains or can form antigens not shared by the host, antibodies are induced which react with the graft antigens so that the graft is cast off or resorbed. The publication of Medawar should be consulted on this topic (78).

Ample evidence exists from the extensive work of the Little group (79) that the ability of tumor cells to survive in the host is dependent on *both* their constitution and that of the host. A tumor cell graft obeys, in part, the rules which apply to a graft of normal tissue. If the tumor carries one or more dominant genes which produce antigens and the host has no, or inactive, alleles of these genes, the tumor graft induces antibody formation against itself. Regression can result under these circumstances. It is known, moreover, that transplantable tumors may mutate so that fewer dominant genes are required in the host for successful transplantation. This is, of course, the same thing as saying that there are fewer dominant genes in the transplant capable of setting up antibody formation. Some mouse tumors, for example sarcoma 180, are almost without strain specificity, and further immunologic study of this strain is desirable. Other mouse neoplasms are exceedingly sensitive to genetic differences, particularly certain strains of mouse leukemia (Furth (80)). MacDowell's (81) consideration of the immunologic aspects of this disease has been particularly detailed. He advances evidence that as soon as a leukemic cell is transplanted it becomes antigenically altered. Homozygous animals can be rendered immune to leukemia of their own line, but this does not serve to prevent the development of the spontaneous disease in the same animals. This concept, however, does not take into consideration the possibility that when the spontaneous disorder occurs the host may be altered in some way so as to be no longer capable of immune response.

It is interesting to consider a correlation between decreasing numbers of dominant genes and increased transplantability. It almost seems as if the cell which has the *least* number of distinguishing characteristics is the cell which is the most virulent, most capable of surviving under circumstances set up by the body as adverse ones, and, perhaps the most malignant. This is consistent with the thesis that the neoplastic cell is a dedifferentiated one. Proof of this requires evidence, not now available, that normal differentiated cells are immunologically much more complex than the primitive cell, and both more so than the neoplastic cell.

The studies of Kidd (11), which are referred to later in detail, are important in their bearing on the question of host resistance to transplanted tumors. They contain an admirable review of the literature on the subject. Attention is called to the fact that some mechanism of protection against transplants exists other than the one characterized by the presence of humoral antibodies. Kidd shows that whereas complement-fixing substances are frequently associated with regression of the Brown-Pearce tumor, and injure the cells of that tissue, similarly specific antibodies have no effect on the course of the V_2 carcinoma. Furthermore, examples are presented of solid immunity in the apparent total absence of such antisera. Barrett (82) has published from Murphy's laboratory beautiful experiments which prove the difficulty, if not impossibility, of creating an immune state to cells homozygous with the host.

In this respect an immunologic study of the differences between the wild strain of Neurospora and the mutant derivatives would be of very great interest. It is not impossible that the more biochemically competent wild organism possesses fewer points of immunologic specificity since there is less tendency to accumulate the products of incomplete or abnormal metabolic processes. Were this concept, as presented by Lederberg (83), to be supported by experimental evidence, the cancer cell could be considered as analogous to the wild strain Neurospora, a back mutation to a completely competent form. It would then become necessary to view the normal cell as the mutant, and again differentiation would be seen as the result of a progressive series of mutations with loss of function and thereby the discovery of special functions.

Abnormal Growth due to Hybridization

An important experiment in the induction of neoplasms is the work on hybridization. Gordon (84) studied species of two genera of tropical fish and found that hybrids between them resulted in the production of melanotic neoplasms in greater numbers than would occur in the parent species. Whitaker (85) described tumors which occurred in the F_1 hybrids of crosses of Nicotiana glauca and Nicotiana Langsdorfi. He ascribed this result to the introduction of chromosomes of the latter into the cytoplasm of the former. Little (86) in a most significant experiment crossed two species of mice, Mus bactrianus and C57 black derived from Mus musculus. The two species differ markedly in size, fertility, and rate of growth. The incidence of tumors in the F_1 hybrids was far greater than could be accounted for even by the addition of the spontaneous tumor-forming tendencies of the parent strains.

It is curious that this type of experiment has not been made more frequently, since it is wholly in keeping with other evidence which has accumulated concerning the development of neoplasms. The work with the Shope papilloma virus in the induction of tumors in domestic rabbits clearly involves the introduction of a cellular protein constituent from one species into the cells of another and not too closely related species, with resultant tumor production and disappearance of the virus. The studies of Claude (87) concerning the enzymatic activities of certain cytoplasmic granules suggest the possibility that such protein particles might modify genes if introduced into the cytoplasm of a cell of another species. Since the genes produce enzymes (25), and modification of enzymes can alter the source genes (51), and since one group of chemical carcinogens form in their metabolism substances toxic to enzymatic activity (69), the route of both carcinogen and filterable factor may be to the gene through the enzyme. This type of thought must be considered as the wildest speculation, however, until much more evidence is available. One would like to have precise data on the ability of enzymes from one group of cells to affect the activity of enzymes from another group, clear proof of a change from normal of the enzyme activities of the cancer cell, more facts bearing on the participation of the filterable tumor-producing agents in the activity of kinetically active chemical systems, and finally a satisfactory demonstration of the ability to alter genes by the introduction of different enzymes into a cell.

The increased rate of tumor production by hybridization, like that by carcinogenic chemicals in genetically suitable material (88), provides another example of gene mutation caused by procedures also effective in causing neoplastic disease. Sturtevant (21) proved that the offspring of the back-crosses from hybrids between two races, A and B, of Drosophila pseudoobscura show a very large increase in mutation frequency. In this material 9% lethals and 0.5% of sex-linked viables were produced without any evidence of increase in the number of major chromosome aberrations like those resulting from x-radiation. This experiment may be of great significance because it proves that mutations can be caused without obvious morphological alteration of the genetically active elements. One argument against cancer as a mutation is the "recognized irregularity of the chromosome equipment in cancer cells." Whereas in given tumors no nuclear abnormality may be discernible, other types present the appearance of extreme chromosome heterogeneity.

Abnormal Growth due to Cell Free Materials

The concept of cancer as a disease caused by a protein component of cells, a "virus," has received increasing attention during the past few years. Indeed, the point has been reached at which so much attention is given to the filtrate or virus etiology as to obscure somewhat the exact sense in which the term is used and its application to the possible prevention or control of cancer.

The original proof of cancer induction by a cellular component came with the demonstration by Rous (89) that a malignant mesoblastic tumor of fowls, occurring spontaneously, could be transmitted to normal birds by the inoculation of cell-free antigenic material from the original growth. Subsequent studies by Claude revealed that the active material is a complex composed of lipo-protein and a ribose nucleic acid, and that it is chemically and immunologically indistinguishable from certain constituents of normal embryonic fowl cells. The conclusiveness of the evidence suggested that other, indeed perhaps all, neoplasms are due to cell-free agents and that cancer in general is an infectious contagious disease. This inclusive and far-reaching view was supported by the fact that a number of other fowl tumors were proved to be capable of being transmitted by cell-free material. Particularly striking is the evidence of Ellermann (91) concerning fowl leukosis. This neoplasm not only can be transmitted by filtrates but also appears to be truly contagious, since contact transmission seems to have been proven completely by the distinguished work of the Michigan investigators (92). Later studies revealed a number of additional examples of active tumor-producing cell-free materials. The Shope papilloma of cottontail rabbits, the oral papilloma of dogs, and the mammary tumor inciter of mice are all agents which may be dissociated from whole cells and which certainly cause benign neoplasms to appear.

The studies of Berry (93) are important in any consideration of the virus etiology of neoplastic disease. This investigator used the principles effective in demonstrating the transforming principle of pneumococci in the transformation of the virus of rabbit fibromatosis to an agent capable of inducing in rabbits a wholly different cytological structure, that of infectious myxomatosis. The material which causes the transformation appears to reside in the myxoma virus nucleoprotein. This mutability of a virus is a striking and frequently observed property. The studies of Duran-Reynals (94) on the Rous agent prove that it has wholly different pathological effects with the production of acute necrotizing lesions, when injected into very young chicks, than it does when inoculated into adult fowls with the resultant tumor of avian mesenchyme. Indeed, this investigator has raised the question as to whether tumor production by this agent may not be a special reaction given by a partly immune host. The fact should be recalled, however, that the Rous agent is quite unique among viruses. It is neutralized by rabbit antiserum to normal fowl tissues and the antibodies of such serum are adsorbed out by extracts of normal chick embryo. Furthermore, chemical studies on the agent reveal a close similarity with a substance obtained from normal chick embryo. The contributions of Claude (95) should be consulted on this subject.

The results reported by Rous and his associates (96) prove that the cutaneous papillomas experimentally produced by the inoculation into domestic rabbits of a cell-free material, the virus of Shope (97), obtained from growths of western cottontail rabbits, frequently result in the development of cancer in the involved tissue. An extensive series of publications has appeared concerned with this experimental study and with certain correlative data derived from immunologic studies with the Brown-Pearce transplantable tumor of rabbits. The conclusion was reached that both types of rabbit cancer investigated are caused by "viruses" which cannot be recovered from the cancer tissue but actually persist in a "masked" form. This is a most important conclusion, if supported by the evidence, since it would prove that at least two and possibly all forms of cancer in mammals are infectious, and, as such, to be transmitted by contagion, even though no disease-producing material can be demonstrated in them.

Morphological evidence has been advanced to prove that the cancers associated with transmissible papillomas in both wild and domestic rabbits arise directly from the papilloma cells. This conclusion is based principally on the finding that the virus causes the differentiating epidermal cells to become much larger than normal and to have much larger vesiculated nuclei with marginated chromatin. It is stated that "all these changes may be encountered individually in tar cancers (for which no virus etiology is claimed) yet when found together they are highly characteristic of the action of the virus" (98). This differentiation by cytological criteria could now conceivably be strengthened by evidence derived from studies employing tracer elements.

Irrespective of whether virus can be recovered from the papillomas or not, an antibody capable of fixing complement, in the presence of a saline extract of the papilloma as an antigen, appears in the blood of rabbits bearing the growths, and it increases in titer as the growths enlarge. Similarly, extracts of papillomas which yield no virus will, when injected intra-peritoneally into normal rabbits, evoke specific antiviral antibody, which fixes complement in the presence of the original antigen and only that material, and, when mixed with virus, reduces or removes (neutralizes) that agent. Not only are the papillomas antigenically active, but the transplanted cancers are also, even though no virus can be demonstrated in them. The virus-neutralizing antibody which circulates in the blood of rabbits carrying the virus-induced papillomas has no effect upon the course of the lesions however, or upon the growth of the cancers, a fact which is ascribed to a protection of the virus by its host cells from the action of the antibody.

The absence of virus from the papillomas and cancers of domestic rabbits is notable, as is its lack in the cancers of the cottontails, although it is abundant in the papillomas of that species. Kidd (99) sees two possible explanations of these findings: that the virus had disappeared from the papilloma when the cancer developed and did not exist in the cancer, or that in both the tissues virus was present but was masked, i.e. could not be demonstrated, though actually present. To prove that the second explanation was the correct one, a search was instituted for a virus-neutralizing substance or inhibitor in the cancer. Table II of Kidd's paper (99) presents the evidence that such an inhibitor was found. Extracts of glycerinated cancer tissue were mixed with papilloma virus and inoculated into domestic rabbits. The controls were of virus mixed with Tyrode solution. No controls with benign papilloma tissue were used, since it was assumed that they could not contain an inhibitor because virus could be demonstrated in them, although the assumption that the cancer did not contain a virus was not accepted. Of fourteen tests with the virus cancer tissue mixtures, virus activity was found in thirteen, but delay in papilloma growth as compared with the Tyrode control was found in eleven. It should be noted that a second control reported, with Tyrode and virus, showed no more activity than was seen in eleven of the fourteen tests presented to prove the presence of an inhibitor (100).

The presence of the inhibitor in tumor is invoked to explain the inability to recover virus from mixed papillomas and cancers of cottontails, as well as from the cancer tissue alone. Friedewald (101) reported that antibody to the papilloma virus is present in extracts of the virus-induced growths of domestic rabbits, even though little or none may be present in the serum. This fact is used to explain the failure to obtain virus from the papillomas and also to explain, as due to extravasated antibody, the presence of inhibitor in the cancer in the cottontail arising in the virus-induced papilloma, a view in support of which morphological evidence of hemorrhage and extravasation is advanced.

Since no way was thought feasible of separating extravasated antibody from the papilloma virus and so of demonstrating the masked virus in the cancer derived from the papilloma of the cottontail, an experiment was made to recover virus from a cancer not caused by it, and so not productive of antibody (100a). Virus was mixed with the tissue of tar-induced cancer of cottontails and the tissue transplanted into normal hosts. Antibodies to the virus were developed and the virus is stated to have increased the speed of growth of the implanted tumors and sometimes to have altered their morphology so that they looked, not like tar cancers, but like virus cancers. From these transplanted virus-infected tar cancers virus could sometimes be recovered if the anti-viral activity of the host's blood had not risen to levels too high to allow a satisfactory test. It is not clear how virus escapes from the cells bearing it to act as an antigen, but the antibody fails to penetrate the cells and kill the virus.

Whereas the papilloma caused by virus inoculation in the cottontail yields virus regularly on test *and* is associated with virus-neutralizing humoral antibodies, in domestic rabbits from flourishing papillomas caused by the same material virus was obtained only irregularly. Furthermore, the sera of the animals that carried growths yielding little or no virus had little or no virus-neutralizing capacity. These findings are summed up by the statement that the facts make it improbable that antibody is responsible for masking the virus in the growths produced by it in domestic rabbits, although the presence of antibody as inhibitor in tumor tissue is invoked to explain the inability to recover virus from the inflamed papilloma cancers of cottontails.

It is remarkable that no antibodies can be shown to be associated with the papilloma, which yields no virus but gives rise to a cancer also not yielding virus but actively productive of antibody. Antibody cannot explain the failure to obtain virus from papilloma of domestic rabbits but is invoked to explain the failure to do so from the cancer of cottontails.

The conclusions drawn from the immunological evidence presented concerning the papilloma virus and the cancer supervening upon it in rabbits is that "the cancers result from virus variation, this in many instances being slight" (100). This conclusion is arrived at because of evidence for the complete specificity for the virus of the antibodies associated with the cancer, and the complete failure to derive any infectious material whatever from the cancers following virus-induced papillomas which are supposedly due to the virus. If the serological evidence for the virus etiology of the cancer is to be accepted, then the virus should be the one reacting specifically with the antibodies, and not a modified one. If the evidence is not to be accepted, it is hard to support the view that the papilloma virus is the etiologic agent. If some modification of the virus is to be invoked as etiologic, then evidence in support of the existence of a modified agent is required. This is advanced in a recent communication by Kidd (11) announcing the demonstration in the V_2 carcinoma of an antigenically specific protein wholly distinct immunologically from the papilloma virus. Further studies will be of interest as they concern the place of the virus in this picture.

Support for the theme of a virus, or extraneous, infectious parasitic etiology for the cancer in rabbits following the virus-induced papilloma is further sought in immunologic studies of the Brown-Pearce transplantable cancer in the same species (10). There was obtained from this tumor a serologically active substance, demonstrable by its ability to fix complement (no agglutination or precipitin reaction was noted) which was insoluble in alcohol, was inactivated by heating to 65° for 30 minutes, by treating with acid to *p*H 4.5 or lower or alkali to *p*H 11.5 or higher, with a uniform particle size over 348μ, and centrifuged down by 20,000 r.p.m. for one hour. These general characteristics of a large protein molecule with the ability to fix complement in the presence of a specific antibody are advanced to favor the infectious etiology of the tumor, though no infectious agent was obtained from it.

The matter of immunologic specificity of neoplastic cells is one of great interest and importance. Beyond doubt new methods and new experiments are desirable. The studies of Andrewes (102) and of Fouldes (103) should be consulted in this regard. These investigators proved that substances related antigenically to the chicken-tumor agents are present in certain fowl tumors elicited with carcinogenic chemicals. The latter have not been proved to be filterable, however, and hence the observation tells us only that the induction of cancer by a carcinogen evokes alterations in avian tissue of molecular structures related in some degree to those which occur spontaneously and transmit the disease.

Cytoplasmic Control of Inherited Cellular Traits

Since the early work in cytology of Roux and Weismann, evidence has been available that constituents of cytoplasm play some role in the development of the cell. The original experiments prove that removal of particular regions of the oöplasm of the unsegmented egg without injury to or disturbance of the nucleus is followed by corresponding defects in the resultant embryos and larvae. This was taken to prove the definite prelocalization in the cytoplasm of the unsegmented egg and specification of the blastomeres due to the nature of the specific cytoplasmic materials which they receive during cleavage. This view was further supported by the classic studies of the Spemann school on the organizer substances. More recently, Sturtevant (104) examined the evidence on the inheritance of dextral and sinistral coiling in the gastropods and found that the eggs of a sinistral individual produce sinistral offspring even if fertilized by sperm carrying dextral dominant factor. J. A. Moore (105) showed that the cleavage rate of echinoderms is a function of the cytoplasm, and E. B. Harvey (106) advanced evidence that in the parthenogenetic merogony of enucleated egg fragments chromatin plays only a minor role in early development.

The experiments of Lindegren, previously mentioned, on the melibiose-splitting enzyme in yeast formed under the control of a specific gene, are important. Once formed, the enzyme is produced as long as the substrate is present, even in the absence of the active allele of the gene which produced it. Beadle (25) draws an interesting analogy between this observation and the formation of pepsin from pepsinogen.

Michaelis (107) studied the plant Epilobium and showed that reciprocal crosses between species and different races of single species show differences between hybrids of first generation and behavior on backcross, indicating that cytoplasmic entities transmitted through the egg are involved and depend upon the nuclear constitution of the plant in which they are found. What is normal in plants of one genetic constitution is abnormal in those of another.

Kühn and Plagge (108) showed that the eggs of Ephestia, homozygous for the *a* allele of the *Aa* gene pair, carry a hormone which requires the *A* allele for its production and which disappears in subsequent generations.

The studies of Imai (109) and of Rhoades (110) on cytoplasmic control have attracted great attention in recent years. In the development of chlorophyll by maize, pure genic control was proved in 100 instances. In two cases, however, cytoplasmic inheritance was shown to exist. In this plant, a specific gene has the power to cause a modification of the plastid marked by smaller size and lack of chlorophyll. The modified plastids have genetic continuity not affected by the constitution of the genes. To quote: ". . . although induced by a nuclear factor, the *ij* gene, the mutated plastid, like a Frankenstein monster, is no longer under the control of its maker."

The work of DuBuy and Woods (111) provides interesting new material on cytoplasmic inheritance in plants and their reports should be consulted directly. They discuss the evidence for regarding plastids as analogous to mitochondria with particular reference to their content of ribose nucleo-protein and to their tendency to break up unless their handling is carried out with great precision. The modified plastids, as the source of spontaneous variegations, are discussed and a virus deemed not likely to be the cause. Of special interest are the data which indicate that certain plant viruses compete with the host chromoprotein for nitrogen and that both normal and abnormal (virus) enzyme systems can be inhibited by cyanide. Success in transplanting variegated plastids with resultant self-propagating abnormalities in host plants are described. The likeness between these studies and the disease induced by the cytoplasmic material of tumor cells induced by the Rous agent is dramatic.

Most interesting of the studies concerning the inheritance of cytoplasmic factors, and perhaps the most pertinent to the cancer problem, are those recently reported by Sonneborn (112). This work is so important that it is discussed *in extenso.*

Sonneborn discovered that one race of *Paramecium aurelia*, when placed in culture with organisms of another race, killed the latter by means of a substance secreted into the culture medium. This ability to kill was assigned to a gene called "killer gene" or *K*. The recessive allele of *K*, which was present in the animals which were killed, was called "sensitive" and designated by *k*. The characters *K* and *k* have been shown to be inherited by simple Mendelian laws, and hence animals can have the genotypes: *KK, Kk, kk.* It was found, however, that the phenotypes, or the expression of these characters, did not necessarily reflect the genotypic make-up. This was due to the elaboration into the cytoplasm by *K* of a substance "Kappa." Kappa is the material that actually determines the phenotype of the organism as to whether it is a killer or a sensitive. *Paramecium aurelia* divides by simple fission, and two sexual processes may intervene. The first of these is conjugation in which two animals come together to form a cytoplasmic bond and exchange, reciprocally, one micro-nucleus. This is preceded by two reduction divisions in each animal. The micro-nuclei, which migrate, fuse with corresponding micro-nuclei in each animal to form a zygote nucleus. There may be, during conjugation, considerable exchange of cytoplasm, a fact which is significant. The second process is autogamy, in which the micro-nuclei in a single animal go through two reduction divisions. Two daughter nuclei fuse to form a zygote nucleus and the others degenerate. Sonneborn found that animals containing the gene *K* and having Kappa present in their cytoplasm are invariably killers. Animals of *kk* genotype which do not have Kappa in the cytoplasm are invariably sensitive. However, the distribution of Kappa in the cytoplasm means that, due to the presence of the cytoplasmic bridge, it can be distributed independently of the micro-nuclei on conjugation. Animals possessing the genotype *Kk* undergoing autogamy might change in genotype from *Kk* to *kk* because of the two reduction divisions and consequent independent genic segregation of these two alleles *K* and *k*. It was possible, therefore, to obtain animals of the following make-up:

NUCLEUS	CYTOPLASM
1. *KK*	Kappa
2. *Kk*	Kappa
3. *kk*	Kappa
4. *KK*	—
5. *Kk*	—
6. *kk*	—

Consequently, animals 1, 2 and 3 were killers regardless of their genotypes; animals 4, 5 and 6 were sensitives even though 4 and 5 contain *K* gene.

It was demonstrated that *K* can maintain and increase the amount of Kappa present in the cytoplasm but cannot initiate its production. Once an animal of composition *KK* or *Kk* is completely deprived of Kappa it remains a sensitive animal regardless. Conversely, an animal of genotype *kk* containing Kappa in the cytoplasm will be a killer although it will revert to sensitive within four to six generations due to dilution of the substance Kappa below the point at which it can manifest the killer character. These animals subsequently will be sensitive.

In examining a number of varieties of killer animals, Sonneborn found, in variety 4, that if grown rapidly with fission occurring quickly, after several generations some of the animals become sensitive. Subsequently, however, these sensitive animals would revert to killer when grown slowly. If he continued to grow them rapidly more and more of them were converted to permanently sensitive animals and fewer of them would revert to killer under the effect of a slower rate of multiplication. In other words, in this case the organisms grow more rapidly than particles of Kappa are elaborated. Sensitive animals result when the concentration of Kappa in the cytoplasm falls below a certain critical level. In the animals which remain permanently sensitive, after the rapid growing of many generations, Kappa becomes entirely diluted out, and even though the genic make-up was *K*, the animals remained sensitive due to the inability of *K* to initiate the production of Kappa.

From these data, Sonneborn was able to calculate the number of particles of Kappa in the normal killer animal, getting an estimate of between one and three hundred particles by two different methods of calculation. He showed that a single particle of Kappa in the cytoplasm, while not enough to manifest itself, was enough to enable the animal ultimately to revert to killer provided the nucleus contained a *K*. His data indicates that Kappa may have a limited power of independent self-reproduction in the cytoplasm in the absence of *K* but that the major factor controlling its elaboration in the cytoplasm is *K*. The number of particles of Kappa necessary to cause the animal to be a killer ranges perhaps from fifty to sixty.

These data indicate that a substance elaborated into the cytoplasm controls the cell phenotype and may be to some extent independent of the genotype. Thus an animal recessive (*kk*) for killer may, in effect, be a killer phenotypically due to the presence of Kappa in the cytoplasm. It indicates also that Kappa is elaborated at a rate independent of the general rate of growth and multiplication. Thus sensitive animals with the genotype *KK* or *Kk* and containing a sub-critical amount of Kappa in the cytoplasm might become killers following autogamy because autogamy prevents fission for many hours and this allows the independent production of Kappa in the cytoplasm to a point at which animals manifest the killer characteristic. Killer animals held for twelve to thirty-six hours at temperatures higher than normal (37°-38°) become sensitive but revert subsequently to killers. This may be due to inhibition of the production of Kappa at higher temperatures, or to its breakdown at these temperatures.

Kappa particles are distributed to the cytoplasm normally in a ratio of 1 particle/600:1000 cu. mm. of cytoplasm.

It should be noted that Kappa in many ways, however, fulfills the requirements of a virus. Sonneborn believes it is not a virus since the factors determining mating types in Paramecia are likewise cytoplasmic, and it is thought doubtful that mating types are determined by virus infections. Clearly, these brilliant experiments deserve extension. It would be most illuminating to have further data on the composition of Kappa, its ability to survive in tissue culture or on chick embryo, and the size of the particles.

The experiments specified concerning cytoplasmic inheritance are discussed, of course, because of the interest in the "virus" theory of cancer, which concerns to a considerable extent the activity of particles supposed to be cytoplasmic in origin and location. The term "virus" is generally employed to specify an agent entirely foreign to the normal tissue it invades, a view derived from the concept of viruses as modified bacteria. It appears that ample evidence is at hand to prove that autopropagating substances which reside in the cytoplasm of cells participate, in certain instances, in the control of inherited characteristics. It is not at all clear and seems unlikely that these cytoplasmic agents are always of cytoplasmic origin and not genic, and subject, at least to some extent, to genic control. The work of Sax (113) bears on this point. He showed clearly that gene products essential for growth are able to diffuse between the cytoplasm of attached cells but not between isolated microspores.

It may be well in this connection to quote Wilson (114): "The chromosomes are as much concerned in the determination of the so-called preformed or cytoplasmic characters as any others." This conservative view is certainly supported by the most convincing evidence and serves to strengthen our contention that somatic mutation, war-weary though the term may be, is still adequate to explain neoplastic disease without the addition of parasitic, foreign, or infectious agents.

Mammary Tumor Inciter (Milk Agent)

One of the most dramatic experimental studies of recent years is the one which has to do with an extrachromosomal factor described by Murray and Little (115) which governs the susceptibility to spontaneous mammary cancer in mice. This was shown by Bittner (116) to be a factor transmitted to the young by the maternal milk. The data on this factor are so numerous and the results so well defined that reference to the original material is required for an adequate consideration of the subject.

The reports of Green (117) have been particularly cogent in invoking studies of the milk agent in support of a virus theory of cancer. The theory is based clearly upon a concept of the disease as due to a "parasite entirely foreign to normal mouse tissue. The origin of viruses from microbes and their nature as highly adapted parasites are reasonable concepts and correspond to the course of parasitism and the nature of highly adapted parasites as observed throughout the biological world."

The milk agent is reported as having the following characteristics: It is present in the milk secreted throughout the lactation period, and in blood and normal tissue as well as in the spontaneous mammary cancer, originally or after serial passages by nursing during the precancerous stage. When passed serially by injection through highly susceptible but not infected mice, the power to cause tumor declined and was lost by the third passage (118). Dilute suspensions are more active than concentrated ones (119). The agent is as active by injection as when given by mouth and resists drying, filtration, lyophilization, and treatment with glycerine. It is heat labile and destroyed by extremes of *p*H. Most of the activity is sedimented by centrifugation at 40,000 r.p.m. (110,000 g.) for one hour. These characteristics, like the ones shown for the antigen in the Brown-Pearce tumor, are the ones of a large protein molecule.

The studies of Andervont and Bryan (120) and of Green (12) show conclusively that the milk agent is active as an antigen when centrifugates of tumor tissue, collected between runs at 15,000 g. and 95,000 g., are injected into rats and rabbits. None of the animals given a mixture of immune serum and agent developed tumors, whereas a large proportion of the controls which received mixtures of the agent with normal mouse tissue antiserum, normal rat serum, or saline did so. These results are interpreted as evidence that "the agent is a virus of exogenous origin."

An equally striking experiment has been reported by Green (13) on the toxic properties of the mouse milk agent antiserum for the mouse mammary cancer cells. Samples of a suspension of mouse mammary cancer cells were mixed with the agent antiserum, normal breast tissue antiserum, normal rabbit serum and saline solution, and injected into susceptible mice. None of the animals which received the mixtures of cancer cells and cancer agent antiserum developed tumor, although nearly all the control animals did so.

The findings of the two studies just discussed indicate beyond doubt that "mouse cancer cells are antigenically different from normal cells." It is also stated that "the same antiserum will neutralize a virus and inactivate the cancer cells stimulated by that virus."

The studies of antibody to the milk agent are similar to the prior work of Kidd (121) on the Brown-Pearce tumor of rabbits, from which no agent causative of cancer can be recovered. Kidd proved that an antibody which appears in the blood of some rabbits implanted with that tumor or injected with cell-free extracts of it is capable of suppressing the growth of the transplanted tumor cells. It was shown, however, that the growth frequently regressed in animals which had failed to develop antibody and, conversely, rare animals showed progressive growth of the tumor in spite of a high titer of antibody. The sera of hosts in which the growth had regressed did not suppress the subsequent growth of the tumor cells by *in vitro* contact unless they contained specific antibody.

The admirable discussion of Kidd (52) should be consulted as a reference review of the immunological factors concerned with neoplasms. He calls attention to the evidence that the fate of transplanted tumor cells is controlled by multiple genetic factors which may decrease during successive transplantation. The usual generalization, previously mentioned, is that the transplanted tissues grow if the hosts have antigens in common with the tumor cells and regress in hosts lacking the antigens and so developing antibodies. The failure to transmit resistance passively and the profound differences between the solid immunity which follows the regression of transplanted tumor and the transient resistance set up by inoculations of tumor material are striking.

It is unfortunate that in the studies of rabbit tumors concerned with specific proteins (viruses) the material is of doubtful genetic homogeneity. The extensive studies of Bittner on the factors which concern the susceptibility of mice to mammary tumors are revealing and should be consulted (122). From these it is concluded that, whereas in the strains of mice studied little or no cancer occurred in the absence of the milk factor, cancer need not occur in its presence unless two other factors, the inherited hormonal influence and the inherited susceptibility, are also present.

Bittner and Huseby (122) have contributed beautifully controlled and most illuminating studies on the factors which bear on the development of mammary cancer in mice. Six lines of animals were used:

1. Group A stock lacking hormone influence. These mice are of low cancer incidence in virgins and high in breeding females. They carry the milk factor as shown by the occurrence of tumors in breeders and transmit it to their young as shown by a 92.5% incidence in nursed F_1 virgins. They clearly have the factor of susceptibility, or there would be no tumors even with the combination of forced breeding and milk influence as shown by the C57 Blacks. The lack of hormone influence is proven by a 3.9% incidence in virgins without the hormone influence provided by breeding. Multiple genetic factors appear to be required to produce the inherited sensitivity to or power to develop the hormone influence.

2. C57 Black stock of low incidence, lacking inherited susceptibility. These animals even when provided with milk factor by foster nursing by A strain females and with hormonal stimulation through forced breeding show only 10.3% of cancer. The absence of inherited susceptibility is apparent. Data obtained from crossing the A stock with the C57 Black animals indicate that the inherited susceptibility is determined by multiple genetic factors, one of which is the gene for brown coat color.

3. Zb and Ax animals of low incidence due to lack of milk factor. These are mice of the high incidence A and Z strains fostered by CBA mothers lacking the factor. Both the inherited susceptibility and hormone influence are shown by a high incidence when the animals are supplied milk factor and force bred.

4. Ax strain virgins of low incidence lacking both hormone stimulation and milk influence. If given the latter only, as by nursing their own mothers, they are, of course, the mice of Group 1, still not liable to cancer until provided with hormone influence by forced breeding.

5. Z strain virgins of high incidence due to the possession of all three factors. AZF_1 hybrids also fall in this group.

6. F_1 hybrids from crossing the A strain lacking hormone influence and Z strain possessing it. A high incidence strain even in the virgin females.

From the data presented in these experiments the fact is apparent that three factors operate in causing breast cancer in mice and any one of the three can be completely determining in its effects. Certainly two of these factors are completely controlled by genes, the inherited susceptibility and the inherited hormone influence. Since from Green's data the milk factor dies out by the third passage through animals not generating it, the likelihood is real that this factor also is gene controlled and indeed has certain similarities to the Kappa factor described for Paramecia by Sonneborn.

Huseby and Bittner (123) provide confirmation for the results just described in their studies of the architecture of the mammary glands in the same strains of animals. Comparisons were made between the three groups each lacking one of the essential factors for the development of breast cancer and with mice possessing all three factors. Precancerous nodules of alveolar hyperplasia were found to occur only in animals with a high incidence of cancer irrespective of which of the essential factors was lacking. It is concluded that "the same three factors that are etiologically important for the development of mammary cancer are necessary for the development of precancerous alveolar hyperplasia."

If a virus is to be defined as an agent which is continually propagated by living cells genetically qualified to manufacture it and which is regularly associated with characteristic properties of those cells, the milk factor and other substances found in tumors certainly qualify. If, however, a virus is to be defined as a parasite entirely foreign to its host tissue, it must meet additional requirements. It should be capable of causing disease in several species, it should be more independent of genetic factors in its pathogenicity, it should be antigenic in the species which is its natural host, it should be capable of being propagated in tissue culture or on the chorio-allantoic membrane of the chick embryo, and the disease caused by it should be contagious. It is clear that some tumor cells contain substances which are in size and composition like viruses and also like genes. One, and the only one clearly etiologic for cancer in mammals, is dependent on genes for its continued production and is equally dependent on genes for its ability to produce its characteristic effects. It is also apparent from the studies of Claude (124) that the protoplasm of mammalian cells, unaffected by viruses or any disorder, contains complexes of protein of similar size and composition. Immunologic studies on these particles would be of great interest, and particularly desirable would be the demonstration of their control by genes or alteration of their composition by substances which are known to change the structure and function of genes, or, particularly, the ability of these particles to alter genes.

There can be no doubt that cells contain substances which affect other cells and can be propagated in turn by the recipients. Ample evidence for the existence of substances of this type has been advanced from work in botany and in embryology. The substances described as neoplastic viruses have not been shown to possess any single characteristic which is not shared by substances which are not infectious parasitic agents. Particle size, antigenicity, protein or even nucleo-protein composition, autocatalytic propagation, inactivation by heat, and changes of *p*H are ubiquitous. Transmission by the maternal cytoplasm or by nursing are all shared by other, non-viral, substances. With all the vast amount of experimentation and highly illuminating, indeed brilliant, results, perhaps transcending in importance those of any other field of biology, the expression coined by Murphy for the Rous agent of "transmissible mutagen" seems to cover the facts more adequately than any other.

Summary

A discussion has been presented of certain mechanisms known to cause self-perpetuating alterations of cells in general. This has been done because of the evidence that neoplastic abnormal growth is due to unique, inherited properties of its constituent cells. It is hoped that from a scrutiny of a variety of data, in part apparently irrelevant to the cancer problem, certain lines of investigation will stand out as potentially profitable in the future.

Evidence has been advanced that the neoplastic process is a distinctive, characteristic sort of abnormal growth, malignant as well as autonomous, something more than a quantitative deviation from the normal rate and extent of differentiation.

Reference has been made to the factors which control normal growth and changes in cells to ascertain to what extent principles normally operative can be invoked to explain the acquisition of abnormal properties. It is clear that the constituents of cells which vary either normally or pathologically are under the control of genes to a very great extent. Gene mutation, either spontaneous or induced, is held to be able to effect the types of changes of cell composition and function, normal and abnormal, which have been described. This principle of genic control is invoked as applicable to self-perpetuating cytoplasmic factors and to maternal influences, as well as to cell constituents known to be the direct product of the gene. The evidence that the ability of substances formed by the organs of internal secretion to affect the growth and differentiation of cells depends upon the inherited properties of the target tissue has been discussed.

Attention has been called to the correlation between the ability of certain agents to cause (a) gene mutations in sexual forms, (b) changes with similar results in vegetative forms or bacteria, not susceptible to study by the conventional methods of the geneticists, and (c) cancer in avian and in mammalian tissue. Whereas this correlation may of course be pure coincidence, it would be surprising if the basic mechanisms in the three types of cells were not similar or perhaps identical.

The induction of certain forms of neoplastic abnormal growth by filtrates of the affected cells, free of the whole cells as such, poses a particularly nice problem to the investigator. In some instances, such as that of the Rous agent, the active material, proven conclusively to be related closely to constituents of the normal cells which it affects, can be recovered regularly from the resultant neoplasm. In other cases, as for example with the cancer which arises on the Shope virus-induced papilloma, this is not true. Here the virus, though etiologically related to the neoplasm, disappears, conceivably incorporated into the affected cells, and a new protein constituent arises, immunologically quite distinct from the inoculated material and from any substance present in the host before inoculation.

Whereas self-perpetuating cytoplasmic entities are well-known in lower forms, and some of these, as the Kappa factor of Paramecia and perhaps certain plant constituents, can be passed from the cytoplasm of one cell to that of another, they tend to be dependent to some extent upon the genes of the host. It would be of great interest if a protein substance could be isolated which, when taken in by the cytoplasm of a cell, could be proved to substitute either for a normal substrate or for a normal enzyme, and as a result would alter the gene to produce a wholly new constituent. Such a compound might disappear and so not be recoverable, but it would have exerted its effect in causing a mutation. There seems to be no fundamental reason why such a mechanism should not exist, and indeed it is wholly possible that this may be one route operative in the production of cancer by physical agents. Dale has shown that x-rays need not affect the gene directly, but may do so indirectly through their effects on enzyme systems. Kensler and Rhoads have proven that the naturally occurring metabolites of one carcinogen affect profoundly certain enzymes of the target tissue.

The studies on the Brown-Pearce tumor have been referred to as providing new evidence for the presence of antigenic constituents in a transplantable neoplasm. The principle is extended by the preparation of specific antibodies against the milk factor in mice and in the V_2 carcinoma of rabbits. There has been little doubt for many years that antibodies can be induced in distantly related hosts by the inoculation of tumor material. From the evidence, however, cancer of spontaneous origin is not productive of antibodies capable of controlling the growth. Some other factor appears to play a role in immunity, and a particularly mysterious and important one. Intensive study is indicated of the antigenic constituents of the neoplasms induced by the conventional physical and chemical agents as well as of those neoplasms known to follow the inoculation of special proteins.

There is believed to be no fundamental incompatibility between the thesis that some cancers are due to self-perpetuating, perhaps cytoplasmic, protein components, and the contention that most are the result of gene change and represent somatic mutations. The dividing line between these two mechanisms has become so indistinct, so much interrelationship between nuclear genes and cytoplasmic constituents has been proved to exist, and such a vast number of different kinds of self-perpetuating entities have been described, that an agent can be found to suit almost any taste.

Some cancers can be transmitted by filtrates and continue to produce the active material, some result from filtrates but yield none of the etiologic substance, some may conceivably result from altered enzymes, either directly, or indirectly through their interaction with genes, and some from direct gene hits with the production of standard mutations. In spite of all these possibilities and the immense number of literary and scientific contributions to the subject, there appears no conclusive proof that any mammalian cancer is caused by an intracellular parasite completely foreign to the host.

Reference has been made to the generalized effects upon cellular differentiation, including the production of cancer, which follow the systemic administration of certain compounds with hormone activity and also of special chemical substances not known to have hormone effects. The possibility has been advanced that perhaps one step in the production of neoplastic abnormal growth may be the chronic exposure of cells to an intolerable environment, with consequent death or defensive variation and the development of protective mechanisms. Such an environment might well allow the variants which occur spontaneously, if they are equipped with protection, to persist and overgrow the normal cells. Since the standard controls of growth are established to govern only standard cells, certain variant cells might not be susceptible to them, and, if so, might grow without restraint. This concept of neoplastic growth will of course be immediately opposed by those who envision the neoplasm only as composed of irreversibly altered constituents which continue to reproduce themselves faithfully forever. Certainly much evidence favors the view of permanent cellular change, but there exist data which require the inclusion of an environmental factor to explain some neoplasms. Furthermore, all cancers need not arise and continue to grow because of exactly the same mechanism. Green (117) has called attention to the fact that two neoplasms, histologically identical, may differ wholly in the degree to which they are autonomous. In man, too many regressions of breast and prostate cancer occur to allow the complete elimination of the general control of cellular environmental factors as a possible participant in the neoplastic process.

From the vast collection of factual material existent on neoplastic growth, one fact in particular stands out as worthy of special mention. This is that further data are urgently required on the comparative compositions of normal and neoplastic cells. These data may concern genes, enzymes, proteins, or any other component. They may be derived from direct analytical methods, from tracer studies, or from the different susceptibilities of the two cell types to toxic agents. Precise information of this type is likely to advance significantly our knowledge of the problem, somewhat more than discussions of whether the etiologic agent is a component foreign to the host or of host origin. It is important to prove only that it is different from any normal component and to learn how to accomplish its destruction.

REFERENCES

1. Papanicolaou, G. N., and H. F. Traut. *Diagnosis of Uterine Cancer by the Vaginal Smear.* Commonwealth Fund, New York, 1943.
2. Brues, A. M., M. M. Tracy, and W. E. Cohn. Nucleic Acids of Rat Liver and Hepatoma: Their Metabolic Turnover in Relation to Growth. *J. Biol. Chem., 155,* 619, 1944.
3. Seidlin, S. N., and L. D. Marinelli. Therapeutic Effect of Radioactive Iodine on Adenocarcinoma of the Thyroid with Functioning Metastases. *Bulletin N.Y. Academy of Medicine, 21,* 440, 1945.
4. Haddow, A. On Secondary Colony Development in Bacteria and an Analogy with Tumour Production in Higher Forms. *Acta de l'Union Internationale contre le cancer, 2,* 376, 1937.
5. Burrows, H. *Biological Action of Sex Hormones.* Cambridge University Press, Cambridge, 1945.
6. Allen, E., and W. U. Gardner. Cancer of Cervix of Uterus in Hybrid Mice Following Long-continued Administration of Estrogen. *Cancer Research, 1,* 359, 1941.

7. Needham, J. Problems of Chemical Embryology. *Fiziol. Zhur., 19,* 241, 1935.
8. Friedewald, W. F., and J. G. Kidd. Induced Antibodies that React in Vitro with Sedimentable Constituents of Normal and Neoplastic Tissue Cells—Presence of the Antibodies in the Blood of Rabbits Carrying Various Transplanted Cancers. *J. Exp. Med., 82,* 21, 1945.
9. Kidd, J. G. Serological Studies in Relation to the Problem of Tumor Causation. *J. Bacteriol., 39,* 349, 1940.
10. Kidd, J. G. A Distinctive Substance Associated with the Brown-Pearce Rabbit Carcinoma. I. Presence and Specificity of the Substance as Determined by Serum Reactions. *J. Exp. Med., 71,* 335, 1940.
11. Kidd, J. G. Distinctive Constituents of Tumor Cells and their Possible Relations to the Phenomena of Autonomy, Anaplasia, and Cancer Causation. *Cold Spring Harbor Symposia on Quantitative Biology, 1947, 11,* 94, 1946.
12. Green, R. G., M. M. Moosey, and J. J. Bittner. Antigenic Character of the Cancer Milk Agent in Mice. *Proc. Soc. Exp. Biol. Med., 61,* 115, 1946.
13. Green, R. G. Cytotoxic Property of Mouse Cancer Antiserum. *Proc. Soc. Exp. Biol. Med., 61,* 113, 1946.
14. Warburg, O. *The Metabolism of Tumors.* (Trans. by F. Dickens.) Constable and Co., Ltd., London, 1930.
15. Greenstein, J. P. *Enzymes in Normal and Neoplastic Animal Tissues.* A.A.A.S. Research Conference on Cancer, The American Association for the Advancement of Science, Washington, D.C., 1945, p. 192.
16. Potter, V. R. *Advances in Enzymology; Biological Energy Transformations and the Cancer Problem.* Interscience Publishers, New York, 1944, pp. 201-250.
17. Roskelly, R. C., N. Mayer, B. N. Horwitt, and W. T. Salter. Studies in Cancer; VII. Enzyme Deficiency in Human & Experimental Cancer. *J. Clin. Inves., 22,* 743, 1943.
18. Haddow, A. Transformation of Cells and Viruses. *Nature, 154,* 194, 1944.
19. Muller, H. J. Induced Mutations in Drosophila. *Cold Spring Harbor Symposia on Quantitative Biology, 1941, 9,* 151.
20. Biesele, J. J. Ribonucleic Acid and Heterochromatin in Epidermal Carcinogenesis. *Cancer Research, 4,* 737, 1944.
21. Sturtevant, A. H. High Mutation Frequency Induced by Hybridization. *Proc. Nat. Acad. Sci., 25,* 308, 1939.
22. Demerec, M. Unstable Genes in Drosophila. *Cold Spring Harbor Symposia on Quantitative Biology, 1941, 9,* 145.
23. Plough, H. H. Spontaneous Mutability in Drosophila. *Cold Spring Harbor Symposia on Quantitative Biology, 1941, 9,* 127.
24. Rhoades, M. M. The Genetic Control of Mutability in Maize. *Cold Spring Harbor Symposia on Quantitative Biology, 1941, 9,* 138.

25. Beadle, G. W. Biochemical Genetics. *Chem. Rev., 37*, 15, 1945 (quote from p. 23).
26. Landsteiner, K. *The Specificity of Serological Reactions*. Charles C. Thomas, Springfield, Illinois, 1936.
27. Wiener, A. S. *Blood Groups and Transfusion*. Charles C. Thomas, Springfield, Illinois, 1943.
28. Tatum, E. L., and G. W. Beadle. Biochemical Genetics of Neurospora. *Ann. Missouri Botanical Garden, 32*, 125, 1945.
29. MacDowell, E. C. *What and Why is Leukemia? A Medico-Biological Study on Mice*. Carnegie Inst. Wash., *Pub. 501*, 353, 1938.
30. Sax, K. Types and Frequencies of Chromosomal Aberrations Induced by X-rays. *Cold Spring Harbor Symposia on Quantitative Biology, 1941, 9*, 93.
31. Glucksmann, A. Quantitative Histological Analysis of Radiation-Effects in Human Carcinomata. *British Med. Bull., 4*, 26, 1946.
32. Gray, C. H., and E. L. Tatum. X-ray Induced Growth Factor Requirements in Bacteria. *Proc. Nat. Acad. Sci., 30*, 404, 1944.
33. Lea, D. E. The Action of Radiations on Viruses and Bacteria. *British Med. Bull., 4*, 24, 1946.
34. Gowen, J. W. Mutation in Drosophila, Bacteria, and Viruses. *Cold Spring Harbor Symposia on Quantitative Biology, 1941, 9*, 187.
35. Dale, W. M., W. J. Meredith, and M. C. K. Tweedie. Mode of Action of Ionizing Radiations on Aqueous Solutions. *Nature, 151*, 280, 1943.
36. Lea, D. E., K. M. Smith, B. Holmes, and R. Markham. Direct and Indirect Actions of Radiation on Viruses and Enzymes. *Parasitology, 36*, 110-118, 1944.
37. Henshaw, P. S. Experimental Roentgen Injury; IV. Effects of Repeated Small Doses of X-rays on Blood Picture, Tissue Morphology and Life Span in Mice. *J. Nat. Cancer Inst., 4*, 513-522, 1944.
38. Furth, J., and M. C. Boon. Liver Changes Associated with a Transplantable Granulosa Cell Carcinoma in Mice. *Proc. Soc. Exp. Biol. Med., 58*, 112, 1945.
39. March, H. C. Leukemia in Radiologists. *Radiology, 43*, 275-278, 1944.
40. Stadler, L. J. The Comparison of Ultraviolet and X-ray Effects on Mutation. *Cold Spring Harbor Symposia on Quantitative Biology, 1941, 9*, 168.
41. Hollaender, A., and C. W. Emmons. Wavelength Dependence of Mutation Production in the Ultraviolet with Special Emphasis on Fungi. *Cold Spring Harbor Symposia on Quantitative Biology, 1941, 9*, 179.
42. Greenstein, J. P., and W. V. Jenrette. Physical Changes in Thymonucleic Acid Induced by Proteins, Salts, Tissue Extracts, and Ultraviolet Irradiation. *Cold Spring Harbor Symposia on Quantitative Biology, 1941, 9*, 236.

43. Stanley, W. M. *Viruses: Currents in Biochemical Research.* Interscience Publishers Inc., New York, 1946, p. 13.
44. MacKenzie, K., and H. J. Muller. Mutation Effects of Ultraviolet Light in Drosophila. *Proc. Roy. Soc.*, Ser. B., *129*, 491, 1940.
45. Beadle, G. W., and E. L. Tatum. Neurospora. II. Methods of Producing and Detecting Mutations Concerned with Nutritional Requirements. *Am. J. Bot., 32*, 678, 1945.
46. Blum, H. Physiological Effects of Sunlight on Man. *Physiol. Rev., 25*, 483-530, 1945.
47. Swanson, C. P. X-ray and Ultraviolet Studies on Pollen Tube Chromosomes. I. The Effect of Ultraviolet (2537A) on X-ray Induced Chromosomal Aberrations. *Genetics, 29*, 61, 1944.
48. Tatum, E. L. Personal communication.
49. Guyer, M. F., and E. A. Smith. Further Studies on Inheritance of Eye Defects Induced in Rabbits. *J. Exp. Zool., 38*, 449, 1924.
50. Sturtevant, A. H., and G. W. Beadle. *An Introduction to Genetics.* W. B. Saunders Company, Philadelphia, 1939.
51. Emerson, S. Induction of Mutations by Antibodies. *Proc. Nat. Acad. Sci., 30*, 179-183, 1944.
52. Kidd, J. G. Suppression of Growth of Brown-Pearce Tumor Cells by a Specific Antibody; With a Consideration of the Nature of the Reacting Cell Constitutent. *J. Exp. Med., 83*, 227-250, 1946.
53. Auerbach, C., M. Y. Ansari, and J. M. Robson. (Unpublished British Work done under military security.)
54. Tatum, E. L. Personal communication.
55. Bawden, F. C., and N. W. Perie. The Effect of HN2 on Tobacco Mosaic and Tomato Bushy Stunt Viruses. *British Military Extra-Mural Research,* June 1944, Z4432.
56. Smith, H. Summary Report on the Systemic Pharmacology and Pathology of the Sulfur and Nitrogen Mustards. *OSRD Report 6325.*
57. Bodenstein, D., and R. Gillette. *Reports of the Medical Division Chemical Warfare Service,* U.S. Army, 1945.
58. Karnofsky, D. A., J. H. Burchenal, R. A. Ormsbee, I. Cornman, and C. P. Rhoads. Experimental Observations on the Use of the Nitrogen Mustards in the Treatment of Neoplastic Disease. *Approaches to Tumor Chemotherapy.* A.A.A.S. Washington. D.C., 1947.
59. Stowell, R. E., and Z. K. Cooper. Relative thymonucleic acid content of human normal epidermis, hyperplastic epidermis and epidermoid carcinomas. *Cancer Research, 5*, 295-301, 1945.
60. Burk, D., and R. J. Winzler. The Biochemistry of Malignant Tissue. *Annual Review of Biochemistry, 13*, 487-532, 1944.
61. Suntzeff, V., and C. Carruthers. The Effect of Methylcholanthrene upon Epidermal Sodium and Calcium. *Cancer Research, 3*, 431-433, 1943.
62. Tatum, E. L. Personal communication.
63. Mottram, J. C. Abnormal Paramecia Produced by Blastogenic Agents

and Their Bearing on the Cancer Problem. *Cancer Research, 1*, 313-323, 1941.

64. Strong, L. C. The Induction of Germinal Mutations by Methylcholanthrene. *Cancer Research, 6*, 501, 1946.
65. Dobriner, K., C. P. Rhoads, and G. I. Lavin. The Spectroscopic Study of Biological Extracts; II. Detection, Isolation and Biological Effects of Metabolites of 1,2,5,6-Dibenzanthracene. *Cancer Research, 2*, 95-107, 1942.
66. Boyland, E., and A. A. Levi. Metabolism of Polycyclic Compounds; III. Anthryl-mercapturic acid. *Biochem. Journal, 30*, 1225-1227, 1936.
67. Jones, R. N., C. E. Dunlap, and C. J. Gogek. The Spectrographic Analysis of Carcinogenic Hydrocarbons & Metabolites; IV. Elimination of 1,2,5,6-dibenzanthracene from the rat. *Cancer Research, 4*, 209-217, 1944.
68. Weigert, F. *22nd Annual Report, British Empire Campaign*, 1945, p. 38.
69. Kensler, C. J., and C. P. Rhoads. Biochemical Studies of Chemical Carcinogenesis, with a Preliminary Note on the Susceptibility to Chemicals of Certain Neoplasms, Both Animal and Human. *A.A.A.S. Research Conference on Cancer*, p. 170. Washington, D.C., 1945.
70. Giese, J. E., J. A. Miller, and C. A. Baumann. The Carcinogenicity of *m'*-Methyl-*p*-Dimethylaminoazobenzene and *p*-mono-Methylaminoazobenzene. *Cancer Research, 5*, 337-340, 1945.
71. Beadle, G. W. *Currents in Biochemical Research; the Gene in Biochemistry*. Interscience Publishers, New York, 1946.
72. Lindegren, C. C., S. Spiegelman, and G. Lindegren. Mendelian Inheritance of Adaptive Enzymes in Yeast. *Proc. Nat. Acad. Sci., 30*, 346, 1944.
73. Dubos, R. J. *The Bacterial Cell in its Relation to Problems of Virulence, Immunity, and Chemotherapy*. Harvard University Press, Cambridge, 1945.
74. Lewis, I. M. Bacterial Variation with Special Reference to Behavior of Some Mutable Strains of Colon Bacteria in Synthetic Media. *J. Bact., 28*, 619, 1934.
75. Schnitzer, R. J., L. J. Camagni, and M. Buck. Resistance of Small Colony Variants (G-forms) of a Staphylococcus Towards the Bacteriostatic Activity of Penicillin. *Proc. Soc. Exp. Biol. Med., 53*, 75, 1943.
76. Griffith, F. The Influence of Immune Serum on the Biological Properties of Pneumococci. *Repts. on Pub. Health Med. Subjects, Gr. Britain*, No. 18, pp. 1-13, 1923.
77. Tatum, E. L. Induced Biochemical Mutations in Bacteria. *Cold Spring Harbor Symposia on Quantitative Biology, 1946, 11*, 278.
78. Medawar, P. B. A Second Study of the Behavior and Fate of Skin Homografts in Rabbits. *J. Anat., 79*, 157, 1945.
79. Little, C. C., and P. A. Garer. The Genetics of Cancer in Mice, In

Genetics of the Mouse. Grüneberg, H., Cambridge University Press, 1943.

80. Furth, J., M. C. Boon, and N. Kaliss. On the Genetic Character of Neoplastic Cells as Determined in Transplantation Experiments. *Cancer Research, 4*, 1, 1944.
81. MacDowell, E. C., J. S. Potter, and M. J. Taylor. The Influence of Transplantation upon Immunological Properties of Leukemic Cells. *Proc. Nat. Acad. Sci., 25*, 416, 1939.
82. Barrett, Morris K. The Influence of Genetic Constitution upon the Induction of Resistance to Transplantable Mouse Tumors. *Jour. Nat. Cancer Inst., 1*, 387, 1940-1941.
83. Lederberg, J. Personal communication to Dr. Stock.
84. Gordon, M., and G. M. Smith. Production of Melanotic Neoplastic Disease in Fishes by Selective Matings; IV. Genetics of Geographical Species Hybrids. *Am. Jour. Cancer, 34*, 543-565, 1938.
85. Whitaker, D. M. Physical Factors of Growth. *Growth Supplement*, 1940, p. 75.
86. Little, C. C. Hybridization and Tumor Formation in Mice. *Proc. Nat. Acad. Sci., 25*, 452-455, 1939.
87. Claude, A., and E. F. Fullam. Electron Microscope Study of Isolated Mitochondria; Method and Preliminary Results. *J. Exp. Med., 81*, 51, 1945.
88. Snell, G. D. *Biology of the Laboratory Mouse*. The Blakiston Co., Philadelphia, 1941.
89. Rous, P. A Sarcoma of the Fowl Transmissible by an Agent Separable from the Tumor Cells. *J. Exp. Med., 13*, 397, 1911.
90. Beadle, G. W., and E. L. Tatum. Neurospora. II. Methods of Producing and Detecting Mutations Concerned with Nutritional Requirements. *Am. J. Botany, 32*, 678, 1945.
91. Ellermann, V. *The Leucosis of Fowls and Leucemia Problems*. Gyldenal, London, 1921.
92. Waters, Nelson F. Personal communication.
93. Berry, G. P., and H. M. Dedrick. A Method for Changing the Virus of Rabbit Fibroma (Shope) into that of Infectious Myxomatosis (Sanorelli). *J. Bact., 31*, 50, 1936.
94. Milford, J. J., and Duran-Reynals. Growth of a Chicken Sarcoma Virus in the Chick Embryo in the Absence of Neoplasia. *Cancer Research, 3*, 578, 1943.
95. Claude, Albert, and Alexandre Rathen. Properties of the Causative Agent of a Chicken Tumor. XIV. Relation between a Tumor Nucleoprotein and the Active Principle. *J. Exp. Med., 71*, 619, 1940.
96. Rous, Peyton, and J. W. Beard. A Virus-Induced Mammalian Growth with the Characters of a Tumor (Shope Rabbit Papilloma). I. The Growth on Implantation within Favorable Hosts. *J. Exp. Med., 60*, 701, 1934. II. Experimental Alterations of the Growth on the Skin: Morphological Considerations: the Phenomena of Retrogres-

sion. *J. Exp. Med., 60,* 723, 1934. III. Further Characters of the Growth: General Discussion. *J. Exp. Med., 60,* 741, 1934. The Progression to Carcinoma of Virus-Induced Rabbit Papillomas (Shope). *J. Exp. Med., 62,* 523, 1935.

97. Shope, R. E. A Transmissible Tumor-like Condition in Rabbits. *J. Exp. Med., 56,* 793, 1932.

98. Kidd, J. G., and P. Rous. The Carcinogenic Effect of a Papilloma Virus on the Tarred Skin of Rabbits. II. Major Factors Determining the Phenomenon: The Manifold Effects of Tarring. *J. Exp. Med., 68,* 529, 1938.

99. Kidd, J. G. The Masking Effect of Extravasated Antibody on the Rabbit Papilloma Virus (Shope). *J. Exp. Med., 70,* 583, 1939.

100. Kidd, J. E. and P. Rous. Cancers Deriving from the Virus Papillomas of Wild Rabbits under Natural Conditions. *J. Exp. Med., 71,* 469, 1940.

101. Friedewald, W. F. Identity of "Inhibitor" and Antibody in Extracts of Virus-Induced Rabbit Papillomas. *Proc. Soc. Exp. Biol. Med., 42,* 330, 1939.

102. Andrewes, C. H. Evidence for the Presence of Virus in a Non-filterable Tar Sarcoma of the Fowl. *J. Path. Bact., 43,* 23, 1936.

103. Fouldes, L. Observations on Non-filterable Fowl Tumors. The Production of Neutralizing Sera against Filtrates of Rous Sarcoma. I. By Non-infective Extracts of a Sarcoma Induced by 1:2:5:6-Dibenzanthracene. *Am. J. Cancer, 31,* 404, 1937.

104. Sturtevant, A. H. Inheritance of Direction of Coiling in Limnaea. *Science, 58,* 269, 1923.

105. Moore, J. A. Developmental Rate of Hybrid Frogs. *J. Exp. Zool., 86,* 405, 1941.

106. Harvey, E. B. Parthenogenetic Merogony or Cleavage without Nuclei in Arbacia Punctulata. *Biol. Bull, 71,* 101, 1936. Development of Half-eggs of Chaetopterus Pergamentaceus with Special Reference to Parthenogenetic Merogony. *Ibid., 76,* 384, 1939.

107. Michaelis, P. I. Über Reziprok Verschiedene Sippen-Bastarde bei Epilobium Hirsutum. *Z. induktive Abstammungs -u. Vererbungslehre, 78,* 187, 1940. II. Über die Konstanz des Plasmons der Sippe Jena. *Ibid., 78,* 223, 1940. III. Über die genischen Grundlagen der im Jena Plasma auftretende Hemmungs-Reihe. *Ibid., 78,* 295, 1940. V. Über die Bedeutung der Genquantität für die Manifestation reziproker Unterschiede. *Ibid., 80,* 429, 1942. VI. In welcher Weise sind an der Manifestation der im Jena-Plasma auftretenden Entwicklungstendenz die Gene dieser Sippe beteiligt? *Ibid., 80,* 454, 1942.

108. Kühn, A. and E. Plagge. Pradetermination der Raupenaugenpigmentierung bei Ephestia kuhniella Z. durch den Genotypus der Mutter und durch arteigene und artfremde Implantate. *Biol. Zentr., 57,* 113, 1937.

109. Imai, Y. A Consideration of Variegation. *Genetics, 13,* 544-562, 1928.

110. Rhoades, M. M. Genic Induction of an Inherited Cytoplasmic Genic Induction of an Inherited Cytoplasmic Difference. *Proc. Nat. Acad. Sci., 29*, 327, 1943.

111. DuBuy, H. G., and Mark W. Woods. A Possible Common Mitochondrial Origin of the Variegational and Virus Diseases in Plants and Cancers in Animals. *A.A.A.S. Research Conference on Cancer,* Washington, D.C., 1945, p. 162.

112. Sonneborn, T. M. Gene and Cytoplasm. I. The Determination and Inheritance of the Killer Character in Variety 4 of Paramecium aurelia. *Proc. Nat. Acad. Sci., 29*, 329, 1943.

113. Sax, K. Diffusion of Gene Products. *Proc. Nat. Acad. Sci., 28*, 303, 1942.

114. Wilson, E. B. *The Cell in Development and Heredity.* 3rd edition, Macmillan, New York, 1925.

115. Murray, W. S., and C. C. Little. Chromosomal and Extrachromosomal Influence in Relation to the Incidence of Mammary Tumors in Mice. *Am. J. Cancer, 37*, 536, 1939.

116. Bittner, John J. Inciting Influences in the Etiology of Mammary Cancer in Mice. *A.A.A.S. Research Conference on Cancer*, Washington, D.C., 1945. p. 63.

117. Green, Robert G. Virus Aspects of Carcinoma. *Minnesota Medicine, 29*, 277, 1946.

118. Green, R. G., M. M. Moosey, and J. J. Bittner. Serial Transmission of the Milk Agent of Mouse Mammary Carcinoma. *Proc. Soc. Exp. Biol. Med., 61*, 362, 1946.

119. Bittner, J. J. Characteristics of the Mammary Tumor Milk Agent in Serial Dilution and Blood Studies. *Proc. Soc. Exp. Biol. Med., 59*, 43, 1945.

120. Andervont, H. B., and W. R. Bryan. Properties of Mouse Mammary-Tumor Agent. *J. Nat. Cancer Inst., 5*, 143-149, 1944.

121. Kidd, J. G. Suppression of Growth of the Brown-Pearce Tumor by a Specific Antibody. *Science, 99*, 348, 1944.

122. Bittner, J. J., and R. A. Huseby. Relationship of the Inherited Susceptibility and the Inherited Hormonal Influence in the Development of Mammary Cancer in Mice. *Cancer Research, 6*, 235, 1946.

123. Huseby, R. A., and J. J. Bittner. A Comparative Morphological Study of the Mammary Glands with Reference to the Known Factors Influencing the Development of Mammary Carcinoma in Mice. *Cancer Research, 6*, 240, 1946.

124. Claude, Albert. Particulate Components of Cytoplasm. *Cold Spring Harbor Symposia on Quantitative Biology, 1941, 9*, 263.

125. Earle, W. R. Production of Malignancy in vitro.
(A) I. Method of Cleaning Glassware. *J. Nat. Cancer Inst., 4*, 131, 1943.
(B) II. Photomicrographic Equipment. *J. Nat. Cancer Inst., 4*, 135, 1943.

(C) Earle, W. R., and L. R. Crisp. III. Microcinematographic Equipment. *J. Nat. Cancer Inst., 4,* 147, 1943.

(D) Earle, W. R., E. L. Schilling, T. H. Stark, N. P. Straus, M. F. Brown, and E. Shelton. IV. The Mouse Fibroblast Cultures and Changes Seen in the Living Cells. *J. Nat. Cancer. Inst., 4,* 165, 1943.

(E) Earle, W. R., A. Nettleship, E. L. Schilling, T. H. Stark, N. P. Straus, M. F. Brown, and E. Shelton. V. Results of Injection of Cultures into Mice. *J. Nat. Cancer Inst., 4,* 213, 1943.

(F) Nettleship, A., W. R. Earle, M. P. Clapp, and E. Shelton. VI. Pathology of Tumors Produced. *J. Nat. Cancer Inst., 4,* 229, 1943.

126. Daniel, G. E., R. R. Spencer, and Dorothy Calnan. Methylcholanthracene and the Environment of Paramecium. *J. Nat. Cancer Inst., 6,* 157, 1945.

X. THE ADRENAL GLAND, A REGULATORY FACTOR

BY C. N. H. LONG[1]

Quel est l'usage des glandes surrénales?
—Academie des Sciences, Bordeaux, 1716

SOME thirty years before the founding of Princeton University, after due consideration of certain unexplored fields of anatomy and physiology, the Academy of Sciences of Bordeaux offered a prize for the best essay submitted on the function of the atrabiliary or suprarenal glands. These organs had been described by Bartholomaeus Eustachius in 1563; but in common with the easy practice of those days, since nothing was known of their function, they were assigned the lowly role of acting merely as supporting elements for their neighboring structures. Even in this regard but little attention was paid then, far less indeed than that lavished on their sister ductless gland, the thyroid, of which it was said by Thomas Wharton in 1656 that its function was "the filling up of the vacant spaces around the larynx, thus contributing much to the beauty and rotundity of the neck, especially in females."

Since the Academy did not see fit to honor any of the contributors with an award, the occasion might have passed unnoted and the record lost to us were it not for the fact that the adjudicator for the Academy was none other than Baron Montesquieu, the noted French jurist, political writer and philosopher. Montesquieu, like many scholars of his day, was equally well versed in the scientific as well as the humanistic knowledge of the age. His skill as a writer and his knowledge of contemporary anatomy and physiology were happily fused in the learned and witty criticism which he later delivered to the Academy (1). In this essay are set forth not only the reasons that prompted the Academy to offer the prize but also those for the rejection of all the contributions that had been submitted.

Let me quote a portion of his preamble in which the reasons for the selection of this subject are set forth. Speaking of the functions of the human body he says, "In this prodigious number of parts, of veins, arteries, lymphatic vessels, cartilages, tendons, muscles, glands, one cannot believe there could be any useless part. Everything contributes to the well-being of the animated subject and if there is any organ

[1] Department of Physiological Chemistry, Yale University Medical School.

whose function is unknown to us we must be attacked by a noble unrest and seek to discover it. It is this which provoked the Academy to select as a subject 'The Function of the Surrenal Glands or Atrabiliary Capsules' and to encourage scientists to work on a problem which, in spite of the researches of so many investigators, was still completely new and seemed to have been so far more the object of their desperation rather than of their understanding."

There is no need to quote all the devastating criticism heaped by Montesquieu upon the unfortunates who had rashly accepted the offer of the Academy to compete for this prize. Two quotations from his criticism will suffice to illustrate what the hopeful contributors to the literature of that day might anticipate from the critics of their papers.

"We have found an author who admits the existence of two kinds of bile; one, the rougher kind, which is manufactured in the liver; the other, more subtle, which is manufactured in the kidneys with the help of the ferment which flows from the capsules through passages of whose existence we are not aware and of whose existence we are even threatened of never being aware. But as the Academy wishes to be enlightened and not discouraged we do not stop at this system.

"Still another describes for us two little vessels which carry the humours from the cavity of the gland into the vein which irrigates it. This humour, which many experiments have led us to believe is alkaline, is used according to this writer to give fluidity to the blood which is coming out of the kidneys after having abandoned there the liquid which composes urine. It must be said that this investigator has only too good supporters for what he proposed: Sylvius, Manget and others had expounded this theory before him. The Academy, which doesn't like duplication and always desires new knowledge, which like an avaricious person who, because of this cupidity wants always to acquire new riches and seems to count as nothing what he already possesses, has not awarded the palm to this system."

Yet Montesquieu could rise above mere destructive criticism. He recognized the difficulty of the question raised by the Academy and the inadequacy of the then existing knowledge of these organs. "Chance [he writes] will perhaps some day achieve what this work could not."

And what has chance and scientific endeavor achieved in the past 230 years? Would Montesquieu now agree that the question posed by the Academy has been answered or would we and our predecessors who have worked on this problem again sit in trepidation as he delivered judgment? While I would hesitate to give a categorical "Yes" to the question asked so long ago by this French Academy, I believe that all, including Montesquieu, would agree that our knowledge of the suprarenal glands is now immeasurably greater, even though it is still far from complete. The chance or chances to which Montesquieu looked for future enlightenment were not soon seized by scientists, for it was not until the period from 1855 to 1867 that another great Frenchman, Claude Bernard, enunciated the two great generalizations that laid not only the foundations of our present knowledge of the ductless glands, but to use his own words established "the basis of general physiology."

The first of these is now known as the doctrine of "internal secretion." In this it was pointed out that certain specialized groups of cells or organs pour their secretions directly into the blood stream and that such organs could thus be termed "ductless glands" to distinguish them from that group whose products passed by ducts to an internal or external surface of the body. Curiously enough Bernard first used the term "internal secretion" in connection with an organ not now generally recognized as an endocrine gland. This was the liver, and he coined the term to describe the capacity of this organ to "secrete" glucose directly into the blood stream. However a few years later he included in this category the adrenals and thyroid glands. To these and certain others the term ductless or endocrine glands is now more generally applied.

Bernard's concept that the blood served as a transporting medium for the products of secretion of certain organs and that these products might be a foodstuff such as glucose or a more subtle agent such as a hormone led in 1867 to the formulation of his general statement concerning the relationship of the blood and body fluids to the preservation of the function or the organism. Bernard recognized that in the higher animals all the cells and organs of the body exist in a fluid environment whose composition is rigidly maintained within limits that are compatible with their existence. The importance of this concept of the constancy of the internal environment to an understanding of the physiology of the higher forms of life was not immediately appreciated. It is only in our own time that it has been given its proper recognition and this is largely due to the work and writings of Cannon, Haldane, and Henderson. Briefly this doctrine states that not only is the composition of the fluid environment of the cells kept within certain narrow limits, but what is equally important, the organism strenuously resists the operation of any factor or circumstance that acts to change this composition.

The physiological processes by which any significant change in the composition of the body fluids is resisted or adjusted to the needs of the moment are often termed compensatory mechanisms. These physiological mechanisms are of many kinds: heat regulation by the skin and lungs, the regulation of water, hydrogen ion and electrolyte concentrations by the kidneys and lungs, and one which is the subject of the present discussion. This is the activity of the endocrine glands which regulate the composition of the fluid environment not only by the action of their products on the excretory organs but also by the effect of these secretions on the chemical processes within the cells themselves.

The factors that tend to distort the fluid composition of the body are not only those imposed by disease, noxious substances and injuries. The requirements of the organism for a continual replenishment of its own material substance demands the ingestion of foodstuffs, minerals, and water itself. It would be an unenviable kind of existence if these had to be ingested at such slow rates that no significant changes occurred in the composition of the fluids both within and without the cells. It is common knowledge that we can ingest at one time considerable portions of our daily requirements of food, minerals, and water without more than temporary dislocations in the composition of our internal environment, so efficient are the bodily mechanisms that maintain this composition.

In addition to this capacity to assimilate at one time considerable quantities of the necessities of life, the higher organisms also possess means to support the level of certain essential constituents of the body fluids at a time when none is being supplied from without. In the absence of such a function all the higher forms of life would have had to remain in close proximity to an ever present food supply and this would have greatly limited the colonization of the land and the possibility of further evolutionary processes.

The vast amount of work carried out in the last fifty years on the function of the endocrine glands has not only served to give an insight into their individual mode of operation but has brought an increasing consciousness of the fact that they act not as independent units but as a correlated and integrated system that is closely associated with the capacity of the organism to adapt itself to the varying conditions of existence.

Thus we find that the endocrine glands not only regulate the metabolism of the organic and inorganic elements of the body but do so in such a manner that their secretions are brought into operation in a way best adapted to the needs of the moment and with consequences that result in the preservation of the internal environment within certain narrow limits. In this regard the adrenal cortex is of particular pertinence, although it is not the only endocrine gland concerned in the regulation of metabolism.

I have been asked to write particularly on the work of my colleagues and myself on the function of the adrenal cortex.[2] What we have done represents only a small part of the many contributions that have been made to this subject in recent years. Though it is quite impossible to mention even briefly all these contributions, anything we have done has been immeasurably helped by the work of our predecessors and contemporaries.

There are two further points I should mention in this general introduction to the subject. The first is that, unknown to Montesquieu and the French Academy, the adrenal glands really compose two endocrine organs, the adrenal medulla and adrenal cortex. This paper is concerned largely with the function of the latter, but it should be emphasized that the methods used and the general hypotheses stated are common to a study of any of the ductless glands. Furthermore the work done on the adrenal cortex has revealed as clearly as in the case of any other gland the integrative nature of the endocrine system.

The investigation of the function of any endocrine gland proceeds along three main lines. These are (a) the chemical nature of its hormone or hormones, (b) the factors that regulate the secretion of hormones by the gland, and (c) the mode of action of the hormone or the manner in which it produces the observed changes in the chemical processes and consequently the degree of activity of the cells upon which it acts.

For my part I have nothing to report regarding the first of these lines of investigation, but I cannot pass unnoticed the brilliant achievements of recent years that have led first to the isolation and more recently to the synthesis of the hormones of the adrenal cortex. For it must be remembered that the isolation in pure form of a hormone is not only an achievement in itself, but makes immeasurably easier the work of those who are investigating the function of the gland from which it comes.

[2] The experiments that are referred to in this paper were carried out in association with Dr. A. White, Dr. and Mrs. G. Sayers, Miss E. Fry of the Department of Physiological Chemistry, and Dr. T. Dougherty of the Department of Anatomy, Yale University.

Chemistry of Adrenal Cortical Hormones

The pioneer studies of Stewart and Rogoff and of Hartman had indicated that various types of adrenal cortical extracts could effect a moderate prolongation of the life of adrenalectomized animals. The recognition that the activity resides primarily in lipid extracts of the gland and that such extracts could indefinitely prolong the life of adrenalectomized animals is to the credit of Professor Swingle of Princeton University and his colleague Dr. Pfiffner. Their methods of extraction furnished the starting material for all subsequent isolations of the active principles.

These active principles were soon recognized to belong to the great class of chemical substances known as steroids. The isolation of these and some twenty others not possessing biological activity is due to the work of Pfiffner, Wintersteiner, and Kendall in this country and to that of Reichstein in Switzerland. The wealth of steroid material present in these glands, both in number of compounds and the diversity of their action, exceeds that found in any other organ of the body, not excluding the gonads, which in their capacity as endocrine glands also secrete steroid hormones. Why the adrenal cortex should contain and presumably manufacture not only the steroids with the characteristic activity of this organ, but also those with gonadal activity, as well as a large number with no known biological activity, is but one of the unsolved problems of its function.

The minute concentrations in which the active steroids of the adrenal, the corticosterones, are present in the gland has made it difficult for investigators to carry out the number of experiments with the pure hormones that they have desired. The generosity of those who have laboriously isolated and then donated these valuable compounds to their colleagues for their work is beyond praise, and as one of the recipients of their kindness I am happy to acknowledge how much it has meant to our work.

It is also a matter of gratification to know that the partial synthesis both of dehydro corticosterone and 11-dehydro-17-hydroxy corticosterone from desoxy cholic acid has now been accomplished. This important work is the result of the collaboration of a number of organic chemists in different parts of the country. The combined efforts of Wallis of Princeton, Kendall of the Mayo Clinic, Gallager at Chicago, Riegel at Northwestern, Wintersteiner at the Squibb Research Institute and Sarett of the Research Laboratories of the Merck Company have resulted in the preparation of quantities of the synthetic hormones far exceeding those ever available from natural sources. The importance of this to the further conduct of research in the field of adrenal physiology is obvious, while the potential value of these new synthetic hormones in the treatment of human disease can now, for the first time, really be explored.

The Regulation of Adrenal Cortical Secretion

The manner in which the activity of the adrenal cortex is adjusted to the needs of the organism and the circumstances which call forth an increased secretion of its hormones is a problem engaging many investigators at the present time. As I mentioned above, one of the tenets of present day endocrinology is that the endocrine glands do not act in an independent and haphazard manner but adjust their secretory rate very closely to the actual needs of the moment.

The second question to be answered is: "What are the factors that regulate the secretory activity of this gland?" This requires for its solution the development of methods that will tell us when the gland is actually producing its hormone at an accelerated rate.

A good deal of information on this problem is now available. In 1930 P. E. Smith (2) showed that removal of the hypophysis in the rat was followed by atrophy of the adrenal cortex and that this could be prevented by the injection of crude pituitary extracts or by implants of the anterior pituitary itself. A large amount of work has since shown that all conditions that cause adrenal cortical enlargement, and these are too numerous to mention, fail to do so in the absence of a factor from the anterior lobe of the pituitary. This factor is a protein hormone now known as the adrenotrophic or corticotrophic hormone. It has recently been isolated from the anterior lobe in a highly purified if not completely pure form by workers in two laboratories. Such a preparation has been obtained from sheep glands by Li, Simpson, and Evans (3), and from hog glands by Sayers, White, and myself (4).

It is interesting to report that these two groups, working independently and with the pituitaries of different species, isolated a substance whose biological, chemical, and physical properties agreed in a remarkable manner.

When injected into animals this trophic hormone produces all the effects given by the cortical hormones themselves. Thus it has been found that it brings about an increased level of liver glycogen and of urinary nitrogen excretion. It causes glycosuria, sodium retention, and increased potassium excretion in rats and has an effect identical with that of the corticosterones in reducing the number of circulating lymphocytes. All these effects can be evoked in the hypophysectomized animal in which no other known agent will cause an increased adrenal cortical secretion. So far as is known at the present time the only effect of this hormone is the stimulation of the secretory function of the adrenal cortex.

But while the identification and isolation of this trophic hormone answers the question of the agent responsible for an increased adrenal cortical secretion, it merely transfers to the anterior lobe of the pituitary the question of the mechanisms that cause this gland to secrete the agent that is essential for an increased hormone production by the adrenal cortex. Are these mechanisms of nervous or humoral origin?

Before such a question can be tested experimentally it is necessary to have available some method of determining the rate of secretion of the adrenal cortical hormones. When we began our work, the available methods for following adrenal cortical secretion were not very suitable for studies in which small animals were to be exposed to a variety of conditions. Those that had been used included (a) the determination of "cortin" excretion in the urine by chemical or biological methods. This is obviously impracticable in small animals, although in man it has been successfully used by Browne and Venning and others to show that trauma or infections increase the secretory rate. The second method (b) used direct determinations of cortical hormone by biological means in the blood of the adrenal vein. This method, applicable only to dogs under anesthesia, had been used by Vogt (5) in England. It has yielded important information, including the demonstration that epinephrine injection is followed by an enhanced output of the hormone. The third group (c) consisted of methods based on the actual content of hormone in the gland. These are not practical at the present time.

Consideration of the question led to the suggestion that some lipid constituent of the gland might vary with different degrees of its activity. It was known that the stainable lipid of the gland decreased under conditions associated with increased secretion, and there was some evidence that the cholesterol content of the gland underwent similar changes. Indeed there was a voluminous literature on this last subject, containing many contradictions and expressions of opinion as to the relation of adrenal cholesterol to the intermediary metabolism of this substance in the rest of the organism. The general idea that cholesterol might be the precursor of the steroid hormones has also been put forth by various investigators but curiously enough no experiments to demonstrate that it varied under conditions of increased secretion of the adrenal cortex or gonads had been done.

We therefore injected the highly purified preparations of adrenotrophic hormone into rats and determined the cholesterol content of the glands after various intervals of time (6). It was found that single injections of this trophic hormone decreased the cholesterol content of the adrenal by some 50% in 3 to 6 hours, while its return to normal levels required many hours. Similar results were observed in hypophysectomized rats, provided too long a time had not elapsed since the operation. The cholesterol content of a variety of other tissues was not affected by such injections. Finally, exposure of animals to such stresses as cold, burns, trauma or hemorrhage brought about similar alterations in the adrenal cholesterol of normal rats but did not alter the level in hypophysectomized animals. Evidently the natural secretion of adrenotrophic hormone following such procedures can also be detected by the changes in the cholesterol content of the adrenal.

In reviewing these results it appeared entirely possible that the rather long period, an hour or more before the adrenal cholesterol fell significantly, might be due to the fact that this substance acted as a reservoir for the continual replenishment of the hormone itself and that other substances in the gland might be more closely related to the actual secretion.

The adrenal cortex is also unique in its high content of ascorbic acid (Vitamin C). Furthermore this gland and the corpus luteum are the only tissues in which such a high concentration of the vitamin is found along with a very high concentration of cholesterol in the ester form. It is unnecessary to remind you that ascorbic acid was originally isolated from the adrenal, although the reasons for its presence in such large amounts were not known. Further study of the literature indicated that as in the case of cholesterol the level of this vitamin in the adrenal was subject to wide variations, it being notable that under conditions of stress low values were found.

Consequently, similar experiments were carried out with adrenotrophic hormone and it was found that the ascorbic acid level of the adrenal fell very promptly after the injection, reaching values of less than fifty per cent of the original level within one hour (7). Indeed, a significant decline could be detected within twenty minutes after injection. Exposure of normal animals to the types of stress mentioned above was followed by marked reductions in adrenal ascorbic acid, yet similar treatment of hypophysectomized animals failed to affect the level of the vitamin in the gland. The ascorbic acid content of other tissues was not affected either by adrenotrophic hormone or by exposure of animals to conditions that were followed by a decline in its content in the adrenal.

While it might be argued that these changes in adrenal cholesterol and ascorbic acid are only an indirect measure of adrenal cortical secretion, our experience has convinced us that they furnish a reliable and rapid indicator of this secretion. As I have pointed out above, this is apparently mediated by the preliminary passage into the blood stream of additional quantities of the pituitary adrenotrophic hormone. Even if these changes in adrenal cholesterol and ascorbic acid were not directly associated with the elaboration and secretion of the cortical hormones but were merely expressions of a general heightening of adrenal metabolism, they would still be of value for following the secretory activity of the gland. However, there is other evidence that more directly links these changes in adrenal chemistry with the actual formation and release of the cortical steroids. Bloch and Rittenberg have shown by the use of cholesterol tagged with deuterium that not only the bile acids but also progesterone are formed from this steroid. The structure of progesterone is sufficiently akin to that of the corticosterones to make it extremely probable that the latter also are formed from cholesterol.

At least one role of ascorbic acid in the adrenal would appear to have been clarified by the recent report of Lowenstein and Zwemer (8) that it is possible to isolate from this gland a water soluble and biologically active steroid in which the steroid nucleus is linked with ascorbic acid. If this is correct then ascorbic acid is actually a part of the hormone in the form it is secreted by the gland, cholesterol presumably furnishing the material from which further quantities of the steroid portion may be elaborated as it is required.

The association of a vitamin with a hormone in this manner is the first known example of a common point of activity of members of these two important groups of substances, although examples of vitamins forming portions of co-enzymes are now fairly numerous. The recent report of Price, Colowick and Cori (9) on the influence of insulin, anterior pituitary, and cortical hormones on the activity of the enzyme hexokinase is an outstanding instance of such interplay, particularly since nicotonic acid amide is a component of the hexokinase system. In this system a hormone, vitamin, and protein enzyme are all part of a cellular mechanism for carbohydrate metabolism.

Finally it may be pointed out that there is an excellent degree of correlation between the decline in adrenal cholesterol and ascorbic acid and such tangible and well-authenticated manifestations of adrenal cortical activity as the increase in liver glycogen in fasting animals and the fall in the number of circulating lymphocytes.

It is therefore possible not only to use these changes in the chemical composition of the adrenal as indicators of increased secretory activity but also under the right conditions to employ them as a method for the assay of the adrenotrophic hormone itself. For this last purpose it is desirable to use hypophysectomized animals as the test objects, since their adrenal cholesterol and ascorbic acid will be affected only by the trophic hormone and not by any nonspecific agent in the material injected. By the use of such a method Dr. and Mrs. Sayers, now at the University of Utah, have found that the blood level of adrenotrophic hormone in normal rats is from 10-15 micrograms per 100 mls. and that a rat pituitary gland contains about 20 micrograms of the hormone. It should now be possible to apply this method to the determination of the level of adrenotrophic hormone in the body fluids, a procedure that should be of particular value in the diagnosis of certain endocrine disorders in man.

We have already made a considerable number of observations on the effect of various procedures on the secretory activity of the adrenal cortex. In most of these experiments we have been satisfied to determine only ascorbic acid, since it reflects any change more rapidly than cholesterol. We have found that exposure to cold (4°C for 1 hour), trauma to muscles and particularly to the long bones, painful stimuli, cutting or bruising the skin, burns, the intraperitoneal injection of cold isotonic saline or other fluids, and the inhalation of volatile anesthetics, all cause an immediate and prompt increased rate of secretion of the cortical hormones (10). As might be expected, the degree of activation of the gland is determined by the intensity and duration of the stimulus applied.

We come now to an interesting and important point concerning the manner in which these diverse types of stimuli bring about an increased degree of cortical secretion. Three recent investigations bear on this point.

(a) We have found that epinephrine, which is certainly released after the most varied types of stress, causes within an hour after subcutaneous or intravenous injection a decrease in adrenal ascorbic acid in normal rats but not in rats 3 days after hypophysectomy. The amounts required are within the physiological range.

(b) Vogt (11) has also found that epinephrine causes a marked increase in the quantity of cortical hormone in the adrenal vein blood of dogs. In chronic experiments, epinephrine produced adrenal hypertrophy only in normal but not in hypophysectomized rats.

In other words Vogt and ourselves are agreed that epinephrine is a potent stimulus to adrenal cortical secretion, although its exact mechanism of action is still unknown.

It is possible to devise more decisive experiments to settle this point, but whatever may be the answer it is of importance that, either directly or indirectly through the adrenotrophic hormone, the stimulation of the sympathetic nervous system and consequent liberation of epinephrine is capable of increasing cortical hormone output. Such sympathetic activity is always a component of those varied circumstances that are known to call forth such secretion by the cortex. Heat, cold, hemorrhage, trauma, burns, toxic agents and painful stimuli of all kinds are always associated with an increased degree of activity of the sympathetic nervous system. Consequently it may be assumed that this is at least a very important if not the only means by which an augmented supply of cortical hormone is released to meet these states of emergency.

I am also aware of the implications of the statement that epinephrine can cause stimulation of the adrenotrophic function of the anterior pituitary. For if it is an agent in this regard then it must act either by the stimulation of adrenergic fibers in the anterior lobe, nerve fibers whose existence is equivocal, or directly upon certain sensitive cells in the gland itself, after the manner in which it accelerates glycogenolysis in skeletal muscle.

(c) A third set of experiments expressing another point of view, or rather another mechanism by which adrenal cortical secretion is regulated, has recently been published by Dr. and Mrs. Sayers (12).

They have found that prior treatment of rats with cortical hormone prevents the usual fall in ascorbic acid when the animals are exposed to cold. We have confirmed this and shown that such treatment also prevents the fall in adrenal ascorbic acid not only after painful stimuli, trauma, etc. but also that produced by epinephrine. I have no doubt that the effect of other types of stress on this component of the adrenal can also be prevented by prior treatment with cortical hormone. Dr. and Mrs.

Sayers suggest that the blood level of cortical hormone determines the secretion of the adrenotrophic hormone. A decline in the former, presumably caused by the stress applied to the animal, acts as a stimulus to the secretion of the latter. If the fall in the blood level of cortical hormone is prevented by injection of exogenous cortical hormone, then the pituitary does not respond. In other words a differential and temporary hypophysectomy is achieved by this means.

These experiments imply that stress produces conditions that cause an increased rate of utilization of cortical hormone in the tissues. This in turn brings about a reduction of the blood level of this hormone which is the exciting factor to the anterior pituitary. Such a view is perhaps half way between that of Vogt and ourselves. It regards the adrenotrophic hormone as the essential factor for cortical secretion but relegates epinephrine to a place with other nonspecific types of stress that reduce the blood level of cortical hormones. Although our experiments indicate that the stimulating effect of epinephrine on the pituitary (if this exists) is abolished by raising the blood level of cortical hormones, it must be remembered that epinephrine itself is a hormone with well-defined points of action and is not in the usual sense "a nonspecific agent." But it is also possible that the well-known metabolic effects of epinephrine may cause an increased utilization of cortical hormone by the tissues or change the composition of the blood passing through the anterior lobe. Only further experimentation can settle these controversial points, but I again repeat that the fact that epinephrine, by whatever mechanism, increases the secretion of cortical hormone is in itself a point of some importance in our understanding of the means by which the organism adapts itself to stress. Indeed this observation furnishes an interesting corollary to the emergency theory of adrenal medullary function put forward by the late W. B. Cannon.

Even with these points unsettled, the picture of the control of adrenal cortical secretion appears at the present time to be as follows:

(a) The changes in adrenal ascorbic acid and cholesterol may be used as indicators of cortical activity. The former is probably directly related to the secretion of the hormone, the latter acting as a reservoir of the precursor of the hormone.

(b) The various circumstances that increase the cortical secretion do so in all probability by first activating the adrenotrophic secretion of the anterior pituitary.

(c) The manner in which anterior lobe activation is effected is either by a decrease in the blood level of cortical hormone or some other unknown changes in the composition of the blood traversing the gland.

(d) The release of epinephrine, which occurs in so many circumstances, increases adrenal cortical secretion. It may do so either (1) by a direct effect on the gland (which is Vogt's view), (2) by stimulation of the anterior pituitary, or (3) as a result of its own effects on the composition of the blood passing through the pituitary.

The series of events just outlined is a good example of the correlation and interplay of three components of the endocrine system, anterior pituitary, adrenal cortex, and medulla. Their combined operation facilitates the adaptation of the organism to external and internal stresses, stresses which if unchecked would threaten its continued existence. Their operation not only corrects changes in the fluid environment that are detrimental to cellular function but also creates conditions within the environment and the cells that are best adapted to meet the emergency.

Mechanism of Action of Adrenal Cortical Hormones

I would now like to turn from a consideration of the regulation of adrenal cortical activity to the equally complicated question of the manner in which the hormones of this gland produce the observed metabolic changes in the cells and body fluids. For it must be in their capacity to facilitate certain types of cellular activity that the protective properties of these hormones reside.

There is not space here to review the large amount of work that has been done on the function of the cortical hormones. I can but remind you that these hormones are particularly active in the regulation of many phases of metabolism. Thus they regulate the sodium, potassium, and water balance by their influence on the rate of kidney tubular excretion and reabsorption of these ions, and in this way they control the shift of fluid and electrolytes between the various compartments of the body. As is now well known, these hormones also have a marked effect on several phases of the organic metabolism. It was recognized many years ago that a subnormal blood glucose level is a frequent finding in adrenal insufficiency in animals and man. Britton has shown that the other carbohydrate constituents of the body, notably the liver and muscle glycogen, are depleted and that the injection of the cortical hormone, even without food, rapidly restores these levels.

In spite of this, it was soon surmised that the control of the electrolyte or carbohydrate levels did not fully explain the action of the hormone. For while it is true that adrenalectomized animals or humans with adrenal insufficiency might die from electrolyte loss or carbohydrate depletion, the mere administration of sodium salts or carbohydrate was not sufficient to return them to a normal state, even though life might be extended by their use.

In 1934 Lukens and I (13) found that adrenalectomy alleviated many of the consequences of total pancreatectomy in the cat and dog. The most notable effect, apart from a prolongation of life, was the reduction in urinary glucose and nitrogen excretion, a fall from the characteristically high to more normal blood glucose levels and a striking abatement of ketosis.

The general conclusion concerning these metabolic changes was that adrenalectomy reduced the high rate of conversion of tissue proteins to glucose (gluconeogenesis) while at the same time allowed a higher rate of carbohydrate utilization. The converse conclusion would be that the adrenal cortical hormones accelerated the conversion of tissue proteins to glucose and/or inhibits carbohydrate utilization by the tissues.

Later experiments demonstrated that in other conditions accompanied by a high rate of gluconeogenesis from protein, adrenalectomy was followed by a comparable reduction in the quantity of protein undergoing catabolism. It has since been shown in several laboratories besides our own that the injection of either the cortical steroids or adrenotrophic hormone into normal fasting animals increases the rate of protein breakdown to a marked degree, while at the same time large quantities of glycogen are deposited in the liver. In fed animals, as Ingle has shown, actual glycosuria follows prolonged injection of either of these hormones, indicating that carbohydrate utilization was also suppressed.

The adrenal cortical hormones, by their capacity to accelerate the catabolic phases of protein metabolism, retard the growth of growing animals. They therefore stand in opposition to the growth promoting factor of the pituitary, which promotes the retention of protein in the body. Since the activities of the adrenal cortex are also regulated by the anterior pituitary, this gland exercises control over the rate of both the catabolism and anabolism of protein.

Though we use such terms as "promotion of gluconeogenesis," "stimulation of protein catabolism," and "inhibition of carbohydrate utilization" to describe the function of the adrenal cortex, we have not yet succeeded in completely defining the action of these hormones. It is true that a stimulation of glucose formation along with a suppression of tissue utilization of carbohydrate is an important factor during periods of fasting. Furthermore the present concept of a dynamic state of protein catabolism indicates the necessity for the regulation of its catabolism as well as anabolism. In spite of all these examples of the teleological significance of the hormone, there is still one feature of adrenalectomized animals that requires consideration before we can accept any present hypothesis as a complete explanation of the function of this gland. This is the remarkable inability of adrenalectomized or hypophysectomized animals to withstand alterations in the external or internal environment that are easily tolerated by intact animals. This intolerance is too well known to need further description, but it should be remembered that merely correcting the disturbances in the sodium or carbohydrate metabolism by administration of these substances is not sufficient in itself to ensure survival.

These measures, while they may bring about the extended survival of adrenalectomized animals living under quiet undisturbed conditions, fail conspicuously to protect such animals under conditions which are easily tolerated by normal animals. Only the cortical hormones can confer a normal resistance to stress to either adrenalectomized or hypophysectomized animals.

This fact and many more that might be enumerated, for example the effect of the hormones on the early fatigue of the skeletal muscles, so well described by Ingle, indicate that there still remains to be found some other basic mechanism of their action. The effects of adrenal cortical insufficiency are extended to many organs, each of which expresses this insufficiency in terms of its inherent function. Thus we observe the rapid fatigue of working muscles, inadequate intestinal absorption, reduced ability of the kidney to separate sodium and potassium, and such defects in liver function as a decreased rate of gluconeogenesis. Yet even with such evidence of the necessity of cortical hormones for the proper function of so many different organs, we must assume until it is proved otherwise that in each organ the point of action of the hormones is the same. At the moment we do not know, except in general terms, what this effect is. All we can say is that without these hormones the capacity of many types of cells to perform their particular function is impaired, and that this becomes particularly evident in times of stress.

Recently some new observations on a well-known effect of cortical hormones have been reported which indicate, if further proof were needed, that we have still much to learn about their action in the body.

Those of you who are familiar with the beautifully illustrated monograph in which Thomas Addison first clearly defined the relationship of the adrenals to the disease in man which now bears his name, will remember that one plate shows the remarkable enlargement of the abdominal lymph nodes in the cases he described.

Since that time, numerous observations both in man and animals have made us aware of the reciprocal relationship between the adrenal cortex and the lymphoid elements of the body. Withdrawal of the hormone is associated with lymphoid enlargement; an excess of hormone causes rapid involution of these elements. This relationship has been particularly emphasized by Selye in his many papers describing the phenomena associated with the "alarm reaction." This investigator has shown that a variety of insults to the organism cause, along with adrenal cortical hypertrophy, a rapid involution of the thymus and lymph nodes.

Dr. White in my laboratory in association with Dr. Dougherty of the Department of Anatomy at Yale have recently made a more detailed analysis of this relationship (14). Their results may be summed up as follows:

(a) The number of circulating lymphocytes is under pituitary-adrenal control. The cortical hormones or the adrenotrophic hormone when injected into normal animals cause a rapid decrease in the blood lymphocytes, reaching a maximum a few hours after injection.

(b) The reduction in the blood lymphocytes is not due to an increased rate of withdrawal by the tissues but is brought about by a rapid lympholysis in the lymphoid tissues themselves. This is further emphasized by the experiments of Reinhart, who found a fifty per cent reduction of lymphocytes in the thoracic duct lymph within twenty minutes after the injection of adrenotrophic hormone.

(c) This lympholysis liberates into the lymph and ultimately into the blood serum the contents of the lymph cells. One of these constituents has been identified as normal serum gamma globulin; a second constituent is probably serum beta globulin. The rise in total serum globulin, due almost entirely to these two globulins, has been shown by electrophoresis.

These experiments indicate an entirely new effect of cortical hormones on protein metabolism. For by this action, intact cellular protein molecules are translocated through the body fluids to all parts of the organism. As in all instances of hormone action this process is not initiated by the hormone but merely altered in the rate at which it proceeds. Once dispersed in this manner this protein must be metabolized by the different tissues in a manner best suited to their requirements, and these presumably vary according to their function.

(d) Since it is well known that the gamma globulins are the fraction of the serum proteins that carry the specific immune bodies, White and Dougherty immunized animals to various antigens, both with and without the injection of cortical hormones. In all cases the titre of the specific antibodies in the serum rose faster and to higher levels in the group receiving the hormone. But this was not all. For if the injection of both hormone and antigen was stopped and sufficient time allowed for the serum titre of antibody largely to disappear, the injection of a single dose of cortical hormone caused antibodies to reappear in a few hours in the serum to levels reached during the height of immunization. Exposure of immunized animals to any stimulus that causes adrenal cortical secretion was followed by similar rapid increases in serum antibodies. In other words this "anamnestic reaction" is a manifestation of the control of lympholysis and serum globulin release by the cortical hormones.

It is apparent that the release of immune globulins in immunized animals is merely a special case of a general phenomenon, for similar quantities of non-immune globulins are released along with other cellular materials from the lymphocytes of non-immunized animals. Yet in the immunized animals the actual release of cortical hormones confers a specific resistance to a particular infecting agent. Is there then a relationship between the lympholysis occurring in non-immunized animals and their resistance to such nonspecific stimuli as heat, cold, trauma, etc.? Is such nonspecific resistance determined at least in part by the capacity of the cortical hormones to liberate from the lymphoid elements into the circulating fluids substances that are vital to the cells for a successful defense against the distortion of their internal environment? This is an unanswered but important question, but these experiments have opened an entirely new line of thought, not only concerning the inability of adrenalectomized or hypophysectomized animals to contend with the factors that distort their internal environment, but also as to the mechanism of action of the cortical hormones. If nothing else, they lend emphasis to the point that there are still unexplored fields in our knowledge of the function of the adrenal cortex.

Such studies are for the future; and since this is so, I too must stand with my predecessors of some 230 years ago before Baron Montesquieu and say that I can only answer in part the question posed by the Academy. I trust that my contemporaries will say as he did to them, "These fruitless efforts are rather a proof of the obscurity of the problem than the sterility of those who studied it."

REFERENCES

1. Biedl, A. *Janus, 15*, 193, 1910.
2. Smith, P. E. *Amer. J. Anat., 45*, 205, 1930. *Anat. Rec., 52*, 191, 1932.
3. Li, C. H., H. M. Evans, and M. E. Simpson. *J. Biol. Chem., 149*, 413, 1943.
4. Sayers, G., A. White, and C. N. H. Long. *J. Biol. Chem., 149*, 425, 1943.
5. Vogt, M. *J. Physiol., 103*, 317, 1944.
6. Sayers, G., M. A. Sayers, E. G. Fry, A. White, and C. N. H. Long. *Yale J. Biol. Med., 16*, 361, 1944.
7. Sayers, G., M. A. Sayers, T. Y. Liang, and C. N. H. Long. *Endocrinology, 37*, 96, 1945.
8. Lowenstein, B. E., and R. L. Zwemer. *Endocrinology, 38*, 63, 1946.
9. Price, W. H., S. P. Colowick, and C. F. Cori. *J. Biol. Chem., 160*, 633, 1945.
10. Long, C. N. H. *Recent Progress in Hormone Research, I*, 99, 1947. Academic Press, New York.
11. Vogt, M. *J. Physiol., 104*, 60, 1945.
12. Sayers, G., and M. A. Sayers. *Endocrinology, 40*, 265, 1947.
13. Long, C. N. H., and F. D. W. Lukens. *J. Exp. Med., 63*, 465, 1936.
14. White, A. Harvey Lectures 1947-48.

NAME INDEX

SUBJECT INDEX